Cambridge Imperial and Post-Colonial Studies

General Editors: **Megan Vaughan**, King's College, Cambridge and **Richard Drayton**, King's College London

This informative series covers the broad span of modern imperial history while also exploring the recent developments in former colonial states where residues of empire can still be found. The books provide in-depth examinations of empires as competing and complementary power structures encouraging the reader to reconsider their understanding of international and world history during recent centuries.

Titles include:

Tony Ballantyne
ORIENTALISM AND RACE
Aryanism in the British Empire

Peter F. Bang and C. A. Bayly (*editors*)
TRIBUTARY EMPIRES IN GLOBAL HISTORY

James Beattie
EMPIRE AND ENVIRONMENTAL ANXIETY, 1800–1920
Health, Aesthetics and Conservation in South Asia and Australasia

Rachel Berger
AYURVEDA MADE MODERN
Political Histories of Indigenous Medicine in North India, 1900–1955

Robert J. Blyth
THE EMPIRE OF THE RAJ
Eastern Africa and the Middle East, 1858–1947

Roy Bridges (*editor*)
IMPERIALISM, DECOLONIZATION AND AFRICA
Studies Presented to John Hargreaves

Rachel K. Bright
CHINESE LABOUR IN SOUTH AFRICA, 1902–10
Race, Violence, and Global Spectacle

Kit Candlin
THE LAST CARIBBEAN FRONTIER, 1795–1815

Hilary M. Carey (*editor*)
EMPIRES OF RELIGION

Nandini Chatterjee
THE MAKING OF INDIAN SECULARISM
Empire, Law and Christianity, 1830–1960

Esme Cleall
MISSIONARY DISCOURSE
Negotiating Difference in the British Empire, c.1840–95

T. J. Cribb (*editor*)
IMAGINED COMMONWEALTH
Cambridge Essays on Commonwealth and International Literature in English

Michael S. Dodson
ORIENTALISM, EMPIRE AND NATIONAL CULTURE
India, 1770–1880

Jost Dülffer and Marc Frey (*editors*)
ELITES AND DECOLONIZATION IN THE TWENTIETH CENTURY

Bronwen Everill
ABOLITION AND EMPIRE IN SIERRA LEONE AND LIBERIA

Ulrike Hillemann
ASIAN EMPIRE AND BRITISH KNOWLEDGE
China and the Networks of British Imperial Expansion

B.D. Hopkins
THE MAKING OF MODERN AFGHANISTAN

Ronald Hyam
BRITAIN'S IMPERIAL CENTURY, 1815–1914: A STUDY OF EMPIRE AND EXPANSION
Third Edition

Iftekhar Iqbal
THE BENGAL DELTA
Ecology, State and Social Change, 1843–1943

Brian Ireland
THE US MILITARY IN HAWAI'I
Colonialism, Memory and Resistance

Robin Jeffrey
POLITICS, WOMEN AND WELL-BEING
How Kerala Became a 'Model'

Gerold Krozewski
MONEY AND THE END OF EMPIRE
British International Economic Policy and the Colonies, 1947–58

Javed Majeed
AUTOBIOGRAPHY, TRAVEL AND POST-NATIONAL IDENTITY

Francine McKenzie
REDEFINING THE BONDS OF COMMONWEALTH 1939–1948
The Politics of Preference

Gabriel Paquette
ENLIGHTENMENT, GOVERNANCE AND REFORM IN SPAIN AND ITS EMPIRE 1759–1808

Sandhya L. Polu
PERCEPTION OF RISK
Policy-Making on Infectious Disease in India 1892–1940

Jennifer Regan-Lefebvre
IRISH AND INDIAN
The Cosmopolitan Politics of Alfred Webb

Ricardo Roque
HEADHUNTING AND COLONIALISM
Anthropology and the Circulation of Human Skulls in the Portuguese Empire, 1870–1930

Jonathan Saha
LAW, DISORDER AND THE COLONIAL STATE
Corruption in Burma c.1900

Michael Silvestri
IRELAND AND INDIA
Nationalism, Empire and Memory

John Singleton and Paul Robertson
ECONOMIC RELATIONS BETWEEN BRITAIN AND AUSTRALASIA 1945–1970

Julia Tischler
LIGHT AND POWER FOR A MULTIRACIAL NATION
The Kariba Dam Scheme in the Central African Federation

Aparna Vaidik
IMPERIAL ANDAMANS
Colonial Encounter and Island History

Jon E. Wilson
THE DOMINATION OF STRANGERS
Modern Governance in Eastern India, 1780–1835

Cambridge Imperial and Post-Colonial Studies
Series Standing Order ISBN 978–0–333–91908–8
(Hardback) 978–0–333–91909–5 (Paperback)
(*outside North America only*)

You can receive future titles in this series as they are published by placing a standing order. Please contact your bookseller or, in case of difficulty, write to us at the address below with your name and address, the title of the series and the ISBN quoted above.

Customer Services Department, Macmillan Distribution Ltd, Houndmills, Basingstoke, Hampshire RG21 6XS, England

Chinese Labour in South Africa, 1902–10

Race, Violence, and Global Spectacle

Rachel K. Bright

© Rachel K. Bright 2013

All rights reserved. No reproduction, copy or transmission of this publication may be made without written permission.

No portion of this publication may be reproduced, copied or transmitted save with written permission or in accordance with the provisions of the Copyright, Designs and Patents Act 1988, or under the terms of any licence permitting limited copying issued by the Copyright Licensing Agency, Saffron House, 6–10 Kirby Street, London EC1N 8TS.

Any person who does any unauthorized act in relation to this publication may be liable to criminal prosecution and civil claims for damages.

The author has asserted her right to be identified as the author of this work in accordance with the Copyright, Designs and Patents Act 1988.

First published 2013 by
PALGRAVE MACMILLAN

Palgrave Macmillan in the UK is an imprint of Macmillan Publishers Limited, registered in England, company number 785998, of Houndmills, Basingstoke, Hampshire RG21 6XS.

Palgrave Macmillan in the US is a division of St Martin's Press LLC,
175 Fifth Avenue, New York, NY 10010.

Palgrave Macmillan is the global academic imprint of the above companies and has companies and representatives throughout the world.

Palgrave® and Macmillan® are registered trademarks in the United States, the United Kingdom, Europe and other countries.

ISBN 978–0–230–30377–5

This book is printed on paper suitable for recycling and made from fully managed and sustained forest sources. Logging, pulping and manufacturing processes are expected to conform to the environmental regulations of the country of origin.

A catalogue record for this book is available from the British Library.

A catalog record for this book is available from the Library of Congress.

Contents

List of Tables   vi
List of Images   vii
Acknowledgements   viii
Acronyms and Abbreviations   x

Introduction   1
1   Chinese Migration and 'White' Networks, c.1850–1902   8
2   The Transvaal Labour 'Problem' and the Chinese Solution   22
3   Greater Britain in South Africa: Colonial Nationalisms and Imperial Networks   38
4   A Question of Honour: Slavery, Sovereignty, and the Legal Framework   70
5   Sex, Violence, and the Chinese: The 1905–06 Moral Panic   95
6   Adapting the Stereotype: Race and Administrative Control   141
7   Political Repercussions: Self-Government Revisited   160
Conclusion: Racialising Empire   185

Appendix: List of Key Figures   191
Notes   197
Bibliography   240
Index   262

Tables

4.1	Numbers of Chinese labourers imported annually	91
4.2	CMLIA Membership: Numbers of Chinese imported per mining group	91
4.3	Distribution of all labour by 'race' on the mines	92
4.4	List of Mines utilising Chinese labour	93
7.1	Repatriations of Chinese labourers at the end of their contracts, 1907–10	175

Images

1	'Twixt the Jew and the Chow' (c.1903)	133
2	Postcard of Chinese labourers (c.1906)	134
3	'The Celestial and his bride' (c.1906)	135
4	Picture reprinted from William Charles Scully (c.1906)	136
5	'A Johannesburg Opium Den' (c.1906)	137
6	'Chinese Labour', British Conservative Party election poster (c.1905)	138
7	'The War's Result: Chinese Slavery', British Liberal Party election poster (c.1905)	139
8	'Chinese or Separation' (c.1906)	140

Acknowledgements

I would like to give special thanks to my PhD supervisor, Andrew Porter, as well as Saul Dubow and Hilary Sapire, whose questions and comments contributed so much to adapting this from my PhD. David Birmingham introduced me to South African history and his encouragement was invaluable. I have drawn many ideas and much support from various academic communities over the last few years. The Postcolonial and Colonial Workshop which Esme Cleall, Emily Manktelow, and I founded at the IHR (Institute of Historical Research) has been a useful intellectual forum, and I wish to thank all attendees. The Global and Imperial History Seminar group at the IHR was equally invaluable in listening to my ideas and encouraging me in so many ways. I would especially like to thank David Killingray, John Stuart, Jon Wilson, Richard Drayton, and Sarah Stockwell.

Several colleagues, friends, and teachers also deserve thanks for their advice: Andrew Dilley, for his enthusiasm and insight into economic history; Robert Bickers, for his immense help with China and Chinese; Anthony Howe, for his guidance on British politics; Esme Cleall, for all those chats about complicated theories and other matters; Kent Fedorowich, for his encouragement and stimulation concerning the British World. I also need to thank Joanna Lewis, Larry Butler, Vincent Kuitenbrouwer, Ian Phimister, Jonathan Saha, Anna Gust, Laura Ishiguro, Karen Hunt, Kelcey Swain, and Frank Bongiorno. Michael Galt will always have my immense gratitude for his support over the years. I could not have even attempted such a global perspective if not for the reading suggestions, translation skills, and general advice of these people. I also feel a great deal of gratitude to my students at LSE, Goldsmith's and Keele; their enthusiasm has so often inspired me to re-examine key aspects of this work.

It is also essential to thank King's College London, the University of London Central Research Fund, and the Royal Historical Society Postgraduate Research Funding for financing my doctoral research trips to South Africa and Oxford, in addition to Dutch-language classes. I also wish to thank the staff at the British Library, the British National Archives in Kew, Rhodes House and the Bodleian Library, the Chamber of Mines, Johannesburg, and all the libraries within the University of London, UK, especially at the School of Oriental and African Studies. In particular, I must thank the friendly and helpful staff at the British Library Newspaper Archives, the National Archives of South Africa in Pretoria, the Library of South Africa in Cape Town, and the fulsome access I was given to the private archives of Barlow World in Johannesburg.

None of this would have been possible without the support of my family. Thank you.

Keele, 2013

Acronyms and Abbreviations

AEJ	*Amalgamated Engineers Journal*
APO	African Political Organisation
Chamber	Transvaal Chamber of Mines
CMLIA	Chamber of Mines Labour Importation Agency
CO	Colonial Office
Corner House	the Johannesburg offices of Eckstein & Co.
ERPM	East Rand Proprietary Mines
FLD	Foreign Labour Department
FO	Foreign Office
HE	Herbert Eckstein Papers
HMG	His Majesty's Government
ISAA	Imperial South Africa Association
LRC	Labour Representation Committee
MP	Milner Papers
NSW	New South Wales
OFS	Orange Free State
ORC	Orange River Colony
PRO	Public Record Office
Rand	Witwatersrand Region of the Transvaal
SAP	South African Party
SAPO	South African Political Organisation
TEA	Transvaal Emigration Agents
TLC	Trades and Labour Council
TMA	Transvaal Miners' Association
UK PP	United Kingdom Parliamentary Papers
Wernher	Wernher, Beit & Co.
WNLA	Witwatersrand Native Labour Association

Introduction

In 1904, the Transvaal began importing Chinese indentured labourers for the Witwatersrand (Rand) gold mines. This was one experiment in a swathe of efforts to rebuild the local economy and pay back some of the colony's crippling war debts to Britain after the South African War (1899–1902). Overall, 63,695 Chinese were imported between 1904 and 1907. By 1910, all of the workers had been forcefully repatriated to China, but their introduction had an important and lasting effect around the world.

Although the Chinese did not remain in the colony, the experimental use of Chinese labour has widely been considered an important episode in South African historiography, and a minor one in the history of the British imperial administration and the 1906 British election. Social and economic historians in particular singled out the reconstruction period of 1902–10 (from the end of the South African War in May 1902 to the Act of Union in 1910 when South Africa became a nation) to track the development of capitalism and labour unionism on the Transvaal, and their relationships with the British administration. Most scholars have argued that the scheme reflected a close partnership between the gold mines and the British government, with the mines being allowed to import cheap, typically powerless, Chinese labour and undermined white and African labourers.[1] Despite George Frederickson writing in the 1980s that 'The traditional Marxist notion that capitalists blind workers to their own interests by cynically playing off one racial group against another does not...do full justice to the normal perceptiveness and intelligence of working-class whites and, at the same time, probably exaggerates the Machiavellian ruthlessness of management',[2] such an interpretation remains dominant. This is partly because the period was largely neglected with the decline in Marxist historiography and the rise of cultural history in the 1990s. To date, the only book solely on the scheme describes it as an example of early industrial globalisation, with economics considered of central import; the cultural and political contexts are largely unexplored.[3] In his writings on the subject, Richardson points out that Chinese importation was an early example of a global market, financed from one continent, production occurring in another, and the workers from yet another.[4] His approach, and that of other writers on the scheme, is excellent at explaining the economic situation to an extent, but the inevitably narrow focus means that the fundamental question of why the Chinese were imported, and the context in which it occurred, have never been fully analysed. Chapters 2, 4, and 6 in particular reveal that the scheme actually brought unprecedented levels of government control and regulation of workers, leading not to a cosy mine–government relations but to fights between the different bodies involved in decision making and between different traditions of knowledge accumulation, especially relating to race. The result was not an all-powerful partnership between the government and mines but a messy affair with competing discourses about race and labour and the role of government in shaping society, which would profoundly shape the economic and labour history of the country, just not as normally described.

The significance of the government's involvement in the scheme goes beyond whether they colluded with the mines; rather, it reflected competing ideas about capitalism and the role of government to regulate or even modify society. An obsession with turning the Transvaal into a 'white' British colony governed the decision process just as much, if not more so, than economic concerns. This cannot only be viewed as a 'process of intracontinental labour mobilization'[5] but as a process based both on the specific colonial situation in southern Africa and the wider history of Chinese indentured labour. Understanding why the 'labour problem' loomed so large in the Transvaal and why this led to the introduction of Chinese labour is an essential part of understanding the whole reconstruction period.

Outside of South African historiography, the subject has likewise been neglected, although it has featured within discussions of the 1906 Liberal landslide or colonial administration under Joseph Chamberlain, Alfred Milner, Lord Selborne and Winston Churchill.[6] Many of these have incorporated a liberal historical tradition, looking to pinpoint in this period 'when things went wrong' in South Africa and whether Britain was complicit.[7] Along this theme, L. M. Thompson's *The Unification of South Africa, 1902–1910* states unequivocally that possible progress was forfeited when the Chinese were brought to the Transvaal. There was a moment after the war, he argues, when white and black unskilled workers had freedom to demand better working conditions and higher wages because of the labour shortage 'and it might have become necessary to modify the labour structure' to secure sufficient labour. However, by importing the Chinese, the dominance of the gold mines over 'the entire South African economy' was cemented, labour and race divisions enshrined in law and Afrikaners were driven further down a hard-line nationalist path.[8] This book, however, argues that this was not a moment when the racial hierarchies in southern Africa could have been overturned if not for mine–government cooperation, as the scheme has often been understood.[9] Such a solution was never seriously considered as both government officials and the mines were a product of the racialised society in which they lived. As Stoler and Cooper have shown: 'systems of production did not just arise out of the impersonal workings of a world economy but out of shifting conceptual apparatuses that made certain kinds of action seem possible, logical, and even inevitable... while others were excluded from the realm of possibility'.[10]

Even with some excellent recent research on Chinese migration globally, this experiment remains understudied despite it arguably being the most thoroughly sourced indentured labour scheme in the world. Only Yap and Man and Harris have looked extensively in recent years at the Chinese in South Africa, although primarily the permanent population there.[11] In other works, the scheme has featured as simply one of many in the general historiographies of indentured or migrant labour worldwide and no detailed analysis of sources has been undertaken.[12] It has been marginal in the historiography of race and empire for similar reasons, although some works have highlighted its important legal implications for immigration within the British empire.[13]

Part of the problem with the more general historiography available is an overreliance on the existing economically focused works which themselves had a tendency to focus too narrowly on a British or white Transvaal perspective of the scheme. They tend to treat the Cape, Natal, the Orange River Colony, and the Transvaal as separate entities, when they were in fact a burgeoning South African nation with overlapping interests in the scheme.[14] This reflects a long-standing assumption within South African historiography that there is no shared South African identity which spans across racial and geographical divides.[15] Instead, emphasis is usually placed on the fragmented geographic or racial identities within the South African state.[16] The historiography that does exist focuses largely on post-1910 nationhood, once South

Africa was already united, exploring Afrikaner nationalism, the torn loyalties of English-speaking residents or opposition to the state from other groups, primarily indigenous Africans.[17] Much of this work has been excellent, innovative, and nuanced, yet it clearly still lives in the shadows of Apartheid. There is an underlying dependency on ethnic classification, partly for ease of writing admittedly, but it can only ever offer a limited, occasionally misleading, perspective of history.[18]

The importance of the scheme is also not limited to southern Africa or to the narrow field of imperial administration. The 'settler' colonies became involved in the controversy, coming as it did right after they had fought in the South African War, to secure it for the British Empire. Consequently, there were many implications for the 'white' colonies in their relationship with each other and with Britain, particularly in matters of self-governance, imperial citizenship, and national and imperial federation. The interest generated led to an unprecedented and unrepeated movement to interfere in each other affairs and was crucial in developing relationships between the settler colonies and Britain, and the too-often neglected relations with each other. As the Conclusion will discuss, it even led to the creation of the Dominions. While imperial and postcolonial scholars have become increasingly interested in dissecting identity politics and networks of people and information, the politics of imperial federation, inter-colonial conferences, and the like are still largely considered minor issues of interest only to elite British colonists in the empire. Provincialism, in Britain and its colonies, is assumed to have been far more prevalent. The subject has been revived, especially thanks to J. G. A. Pocock's writings and the recent 'British World' efforts,[19] but the vagueness of the categorisation and the frequent focus on ethnicity continue to frustrate attempts to fully encompass it within the mainstream of imperial history. The importation of Chinese labourers into South Africa arguably had a greater effect on establishing future relations between these places than anything else apart from the world wars.[20] Through it one can trace the complicated negotiations between national and imperial identities, between independence and patriotism, and giving a clearer sense of how the relationships between these places evolved in relation to each other and, crucially, in relation to non-ethnic Britons.

The application of crucial concepts about colonial knowledge and geographical concepts of networks and webs is an important part of this effort, especially in Chapters 1 and 3 and the Conclusion. By looking at the short time frame from 1902, when the scheme was first seriously considered after the South African War, to the repatriation of the last

Chinese in 1910, this offers a chance to study global history while retaining depth by looking at one particular event. Such an approach can help show 'how historians might connect the study of this global sharing of knowledge and experience to local resolutions of the worldwide problems of race and labor, governance and citizenship, dependence and independence'.[21] It allows an exploration of 'those circuits of knowledge and communication that took other routes than those shaped by the metropole-colony axis alone'.[22] Just as Ballantyne demonstrated regarding the knowledge circuits which spread and adapted versions of Aryanism, looking at Asian migration and the corresponding networks of ideologies and peoples 'reveals the profoundly mobile character of racial knowledge and discourses about cultural difference within the British empire, a reality that necessitates a trans-national analysis [of] imperial knowledge production'.[23] Few historians have embraced the study of what he calls 'horizontal mobility', the forms of movement and cultural traffic that linked colonies in the 'periphery' together. It is still relatively new to think in terms of the colonies having relations with each other.[24] And despite a few impressive studies, most work continues to use the traditional metropole–colony relationship. For instance, both recent Oxford Companion books on Australia and Canada failed to include anything on their relationships with other parts of the empire.[25] Yet the discourses which arose out of the 'yellow peril' reveal this 'horizontal mobility' perfectly; because Britain was largely marginalised, this is an ideal way to study the ways colonies interacted with each other, the development of colonial nationalisms and the roles the United States played within these networks. Understanding the global dimensions of the debates regarding Chinese migration can better help us understand the dynamics of migration and information networks connecting white, English-language communities while 'transgress[ing] the analytical boundaries of both metropolitan-focused imperial history (where the empire is viewed from London out) or histories of individual colonies (where the view is from the colony towards London)'.[26] This is especially important as 'the inadequacy of national "boxes"', or imperial ones, 'would seem to be especially and emphatically true of the "neo-Britons"'.[27] Because imperial federation failed to materialise, and because of the dominance of national histories, these matters are rarely taken seriously. Yet, as Frederick Cooper has argued, if one wishes to analyse history with any rigour, '[e]ven more important is what one does not see: the paths not taken, the dead ends of historical processes, the alternatives that appeared to people in their time'.[28] The regional differences, the ultimate authority of Britain, and the rise of nation states do not negate the monumental importance of these networks during this period. This book shows a process, a series of contested discourses, ephemeral webs of communication, and information flows both particular to the time and yet significant to wider global and imperial structures and cultures.

Looking closely at this one case also allows space for the exploration of specific local issues, such as the analysis of the moral panic which arose in the Transvaal in 1905–06, explored in Chapter 5, or the Cape, British, and Transvaal elections covered in Chapters 3, 4, and 7. This enables a clearer understanding of racial and political developments in southern Africa during the reconstruction period and the formation of South Africa, the nation, while placing these localised issues within a wider historiography, such as research into moral panics and Black Perils or global networks of labourism, humanitarianism, trade, and imperialism.

The first chapter provides a history of Asian migration, especially Chinese, to the white colonies, because the racial stereotypes which formed, and the immigration restrictions implemented, were very important in shaping the development of the scheme and, in turn, the scheme would influence the position of the 'white' colonies and the place of Asian migrants within them. Chapter 2 then answers the question of why the Transvaal imported Chinese labour at all, especially when it was so controversial. Chapter 3 explains how the decision to import Chinese labour affected relations between the 'white' colonies and Britain and how networks of information and the dynamics of power existed at this time. The debate over the scheme led to a unique moment in the history of empire, with the settler colonies debating with Britain over who should ultimately have a share in colonial governance, with important ramifications for imperial federation, responsible government, and the place of non-whites within a democracy.

This book addresses the historical and social construction of identity throughout the settler colonies and Britain, and consequently, much consideration was given to the racial terminology to use in this book, and to the changing names of places. When possible, contemporary usage has been retained. The term 'non-white' in this book is perhaps the most problematic and serious consideration was given to alternatives, such as black, which in this context are equally problematic.[29] The term 'non-white' is imperfect, but does signify the historical realities of white political dominance and the ways that peoples from China, India, Japan, Africa, and elsewhere were labelled in relation to 'whiteness' at the time. Using other terms like 'black' would lose much of the complexity of race and other forms of identification during the period. That said, non-offensive terms have been chosen when not anachronistic. For instance, Afrikaner has primarily been used instead of Boer.

There is perhaps one major failing of the sources consulted for this book. The scheme was almost certainly the best-documented indenture labour scheme ever, but no Chinese language sources have been consulted, opening the way for interesting future research. The sources available in English and Afrikaans, however, are far too numerous to appear in a single book and Chapter 4 attempts to specifically address official Chinese and British responses to the scheme. Dissection of the administrative and legal negotiations reveals to what extent the scheme became an important issue of national honour both in Britain and China at the time. In Britain, this reflected a post-South African War crisis of conscience over the connection between state and capital and about the conflicting ideals of 'white labourism' and the civilising mission. For China, it marked the first time the government directly negotiated over the terms of indenture, in an attempt to address (mainly unsuccessfully) many of the weaknesses of the Qing dynasty at the time, mainly regional power struggles, the growing support for Sun Yat-Sen in the diaspora, and imperialist incursions on China.

Chapter 5 examines the moral panic which arose in 1905 and 1906 as a result of economic and social tensions, fed by networks of sensationalist information regarding the alleged propensity of Chinese men for violence and sex. It also helps situate the yellow peril historiography alongside both African 'Black Peril' literature and theoretical work on moral panics. This also allows an analysis of violence as a 'performative act', symbolising attempts by different groups (usually male) to demonstrate control.[30] Chapter 6 explains the ways the administration and mines tried to adapt the scheme after the public panic through contested forms of colonial knowledge and an increasingly acrimonious power struggle in their own attempt to demonstrate their control over the situation. Chapter 7 then explains the significance of the scheme in shaping the South African nation, through the 1906 British and 1907 Transvaal elections; even the timing of the granting of responsible government and union were largely determined by the scheme. The Conclusion reflects both on the scheme and on the lasting legacies of it for South Africa and the rest of empire, especially the formation of a two-tier British empire based on race.

1
Chinese Migration and 'White' Networks, c.1850–1902

In 1902 the idea that a British administration would organise the migration of tens of thousands of Chinese to a 'white colony' seemed impossible to imagine. The entire controversy over Chinese migration, and why the scheme in South Africa seemed so unlikely, related to the development of overlapping, but often conflicting ideas of democracy, whiteness, and Britishness which developed in the settler colonies and the United States. As Bridge and Fedorowich have explained, 'being British anywhere meant exercising full civil rights within a liberal, pluralistic polity, or at least aspiring to that status. "Whiteness" was a dominant element'.[1] The development of identifications and networks based upon whiteness and Britishness, and the occasional conflict between these identities, largely depended on an African or Asian 'other'. While indigenous peoples were the 'others' of the early nineteenth century,[2] each colony had a distinct, and in most colonies a decreasing, 'native problem', whereas Asian migration increasingly became *the* issue which could unite disparate parts of the settler colonies around a network of exclusionary whiteness. While the African 'other' remained predominant in southern Africa because they made up the majority of the population there, the debate over Chinese labour importation into the Transvaal briefly transcended local racial issues and created a unique inter-colonial dialogue about the relationship between the settler colonies and Britain. It is thus essential to explain the controversy surrounding Asian migration within the burgeoning 'white' colonies in the late nineteenth and early twentieth centuries in order to understand both why the Transvaal Chinese indentured labour scheme was considered and why it became a major global news story.

To understand the issue of Chinese migration during this period, the 'network' metaphor is particularly useful. On the one hand, Britain made the laws and so held a central position in debates. At the same time, Britain was itself largely absent from taking an active role in these debates. I choose to use this term 'network' because specific migration and communication flows led to shared imagery and ideological discourses. 'White', largely English-language, networks of people and information disseminated shared concepts of 'Asians'[3] in southern Africa, Australia, New Zealand, Canada, and the United States. This 'white' network largely transcended class barriers, uniting the working classes with the intellectual and political elites in the self-governing colonies through a shared concern.

This chapter deals with this particular network at a particular time. Subsequent chapters address overlapping yet different networks of ideas and peoples in relation to the Chinese indentured labour scheme in South Africa. Whiteness was not always so dominant in these discourses, nor did Britain always play a marginal role, yet at the same time, networks between these places were often created and sustained precisely because Britain was at the centre, at least in terms of governance.

With these points in mind, this chapter will first explain how Asian indentured labourers came to be seen as desirable while also seeming to threaten 'white' dominance, largely from the 1850s when European and 'free' (unindentured) Asian migration increased to the temperate zones, which were thought most suited to European settlement.[4] It will then chart how this 'white' network reacted to Asian migration and how shared stereotypes were developed and spread throughout the world. The chapter closes by demonstrating how interlinked the issues of imperial federation, burgeoning nationalism and suffrage had become within this network.

Asian migration

In the nineteenth century, about 50 million Europeans, 50 million Chinese and 30 million Indians migrated globally.[5] Of these, almost 750,000 Chinese and 1.5 million Indians were formally indentured to European employers for use in their colonies.[6] Indentured labourers, or 'coolies',[7] were first used by European colonials in the Dutch East Indies in the 1600s. So valued did the Chinese 'coolies' become that within two weeks of arriving at the Cape in 1652, Jan van Riebeeck, in charge of founding a Dutch settlement there, wrote in his diary that he longed for Chinese gardeners to help.[8] In the British colonies, it was only following the 1833 abolition of slavery that plantation owners looked to Asia due to the sizeable and impoverished populations there.[9] The Mauritian and West Indian sugar industries were the first in the British colonies to successfully secure labour from British-controlled areas of India, while Peru and Cuba imported Chinese.[10] As shipping became more organised, cheaper, and faster mid-century, indentured labour was also adopted in places without a history of slavery. Small numbers of Pacific islanders and Chinese were imported into New South Wales (NSW) between 1848 and 1852, to replace convict labour and assist the European wool industry. Natal too began an Indian indenture scheme in the 1860s as one of their first acts as a self-governing colony, to help work their sugar plantations.[11] This labour aided colonial businesses and most settlers accepted it as a commodity necessary for economic success, because the climate allegedly did not allow for whites to do physical labour or because of a general labour shortage.

While most indentured labour was considered desirable, 'free' Asian migration (especially Chinese) to the self-governing colonies was almost universally viewed negatively. 'Free' is a slight misnomer, but essentially meant they operated their own loan systems to organise and control migration, outside the direct control of Europeans.[12] Indentured labourers too could become 'free' once their period of service had ended, and increasing numbers chose to do so. From 1848, a series of gold discoveries in California, British Columbia, and Australia attracted increasing numbers of 'free' Chinese men to work, placing them in direct competition with European settlers. While indentured imports were usually limited to manual rural labour, with little chance to compete directly with white settlers, 'free' migrants went primarily to urban areas. McKeown estimates that 135,000 Chinese migrated to Australia and California in the 1850s alone, all 'funded and arranged by Chinese capital (albeit transported on European ships)'.[13] In 1851, the Chinese government repealed the law which had made migration outside China a criminal offence, punishable by death, further increasing the flow outwards. The Opium Wars of the 1840s and 1860s also strengthened European, especially British, control of Chinese ports from which potential labourers could be recruited. And 'free' migrants tended to follow their indentured fellow nationals, regularly working as traders or artisans either servicing 'coolie' needs or benefiting from the good reputation for work which their indentured predecessors had.[14] It is no wonder that Richardson called the use of such labour in the Transvaal 'one of the most important African episodes in this process of intracontinental labour mobilization'.[15] But such a focus only reveals part of the story. This was not merely a case of cheap labour mobilisation by an increasingly globalised economy or about easier transportation. It was equally about the global movement of ideas.

Asian stereotypes

The long European tradition of stereotyping nations like China and India already existed in the settler colonies before they were a physical reality.[16] Stuart Creighton Miller, for instance, has demonstrated how stereotypes spread in America, gathered from travellers, traders, missionaries, and newspapers, before the Chinese had actually arrived.[17] The predominant view by the mid-nineteenth century regarded 'China as an exotic, backward, only semi-civilized, and in some ways rather barbaric country'.[18] The place and its people were initially objects of curiosity and some ridicule but certainly not a threat, denoted by the Chinese indenture trade being called the 'pig trade' and the Chinese male hairstyle as a 'pigtail'.

When 'free' Chinese started arriving in large numbers in the settler colonies mid-century, pre-existing stereotypes about the Chinese nation were applied and adapted to the 'threat' they posed. Such a process of knowledge exchange and stereotyping was not new, but it did accelerate and harden during this period. And since race was always an unstable identification, it was natural for settlers to develop stereotypes about each other. These shared stereotypes were fostered by a common language and a steady flow of migration and information, especially between the mines.[19] The increasingly international business of book and newspaper publishing and dissemination, growing literacy rates and the improved speed of railway and ship travel only aided this flow.[20] Steamboat services regularly carried news and people along the Pacific and Indian coastlines; border control was almost non-existent. Many of the growing number of colonial newspapers based their international news on 'verbatim reprints from British papers that had arrived by ship'.[21] The spread of Reuters' services around the world also ensured an increasing uniformity in many newspaper's coverage, with English-language newspapers increasingly inserting the same stories from Reuters, which in turn ensured a globalisation of many discourses.[22] The Colonial Office (CO) rarely supplied information to colonies about each other and there simply was no official system for them to communicate with each other, so information networks were largely informal in nature and, in the case of Chinese stereotypes, largely bypassed Britain while incorporating the United States, particularly the West Coast.

One phrase in particular was often repeated over the next 50 years in all these places to sum up Chinese stereotypes: the Chinese were disliked as much for their virtues as for their vices. This duality of fearing and admiring the Chinese reflects Sander S. Gilman's argument that negative and positive stereotypes are always connected. The virtues are 'that which we fear we cannot achieve' and the vices are 'that which we fear to become'.[23] The vices were usually connected to 'immorality', especially sexual. The Chinese were also feared and desired because they were perceived as hard-working and cheap and there were many of them; these were clearly admirable traits, but ones which became points of anxiety. In Sydney, the governor of NSW, Charles Augustus Fitzroy, claimed that the 'free' Chinese there had easily found employment as servants, gardeners, and shepherds 'and have generally proved to be an industrious and harmless class of men – giving satisfaction to their employers'.[24] In British Columbia, their labour was considered so essential to the railways that 5000–6000 Chinese were specifically imported from Hong Kong in 1882, under contract to the Canadian Pacific Railway.[25] As the economist Persia Campbell explained, when an employer 'required a labour force he could go to a Chinese merchant and contract for it'. A Chinese go-between would then supply the required number of workers and if any problems occurred, the 'coolie was removed...[the manager] was always in a position to secure the same number of substitutes'.[26] The seemingly endless supply of Chinese labour and the ease by which one labourer could be substituted for another was an obvious attraction for the growing industrialisation of the colonies, especially when they were often paid lower wages than their European counterparts.

However, it was one thing for 'coolies' to work on plantations, doing work which whites could or would not do, but quite another for them to compete in the more lucrative industries of trade and gold mining. 'The struggle was perceived not simply as between Europeans and Chinese, but between white labour and capitalists using Chinese as their pawns' to lower wages and prevent the spread of labour unions.[27] And because these things were broadly perceived to be happening throughout the self-governing colonies and in the United States, a similar imagery was evoked to describe their situations, 'fuelled by the human rivers of migration' from both Asia and Europe and growing unionisation.[28] When diamonds were discovered in Kimberley in southern Africa, miners there quickly mirrored the sentiments of their counterparts in Australia, California, and elsewhere:

Chinamen are industrious, saving, sober, peaceful people; and they succeed in making a livelihood, and even in acquiring what is to them wealth, where an Englishman would starve. But on the other hand, they rarely or never take root in the country where they settle; they leave their women behind them, and they invariably endeavour to carry out their own laws and social system – by secret means, if necessary – in the place where they have temporarily taken up their abode.[29]

Clearly the network of European labour between the mining communities was important in developing a sense of commonality among the settler colonies and the United States, and in developing a rather standardised view of what Chinese immigration would mean for white settlers. 'It was this conjuncture which created a context in which defining themselves and their labour market interests as "white" could seem an advantageous option to organised workers.'[30] The Chinese were the firm 'other', as much the enemy of working class interests as employers were thought to be.

Exclusionary policy adopted

Once 'free' Chinese migrants were deemed undesirable by mostly working-class and male whites, they sought to establish exclusionary measures.[31] California was the first to implement legal restrictions on Chinese migrants through local mining codes in the 1850s, although illegal expulsions of Chinese from mining communities had begun in 1849, almost as soon as any Chinese had arrived. However, by 1860, the Chinese population in California was 35,000, roughly a quarter of the mining population, and the most clearly 'foreign' looking of the diverse group there.[32] When Victoria and NSW had similar migration patterns during their gold rushes in the 1850s and 1860s, they both adopted anti-Chinese legislation modelled on California's, including poll taxes and limits on the number of Chinese any single ship could import at a time.[33] As Guterl and Skwiot have noted, with only a little exaggeration: 'By the end of this story, no policy was singularly original.'[34]

The appeal of Chinese exclusion was not limited to mining communities, however. If it had been, no exclusionary legislation would have been enacted. In none of the places under discussion were labour unions strong enough to push through legislation on their own. Like their white counterparts, numerous Chinese drifted into factories, railway construction, and farm work (where continual labour shortages and the prevalent misconception that Asian labour was cheaper meant they were always desired by a few). Some even established their own businesses in direct competition with white settlers. With growing numbers of white migrants also arriving and looking for work, wider sections of the white settler population demonstrated antipathy towards Asian migration. The alleged intelligence of Asian populations, their numbers and low standard of living threatened white supremacy in the settler colonies in ways that indigenous populations were rarely seen to do. While employers were largely satisfied, European migrants who felt their jobs were being taken away were not. Companies several times used Chinese men to replace striking Europeans.[35] They also increasingly replaced factory workers in semi-skilled and skilled positions in Australia, the United States, and Canada. For instance, one Nevada union complained in 1869 that 'Capital has decreed that Chinese shall supplant and drive hence the present race of toilers...Can we compete with a barbarous race, devoid of energy and careless of the State's weal?'[36] Senator Perkins of California described the situation as a fundamental question of racial dominance: 'when two races, as radically different as the Chinese and Americans freely intermingled there were only two possible outcomes: assimilation or subjugation...dominate or be dominated'.[37] This, it was thought, posed a very real threat to the traditional European family unit; whites would have to end their family bonds in order to compete. It would be the death of the nuclear family and 'a descent in the scale of civilization' enjoyed by European settlers.[38] While some business people would not mind as long as they made money, the attitude increasingly developed that it was the job of the government to legislate to protect white settlers from the Asian menace. If it did not, democracy would fail.

Of course, such a stance highlighted the inconsistencies of white ideology. The very idea of 'white' colonies depended 'on the premise that multiracial democracy was an impossibility', as Lake and Reynolds have recently shown.[39] But Britishness, and the British Empire, were also supposed to help spread democracy. The idea that racial mixing was anathema to democracy grew out of 'the great tragedy of Radical Reconstruction in the United States' and popularised by 'the British Liberal politician and historian, James Bryce, whose *American Commonwealth* was taken up as a "Bible" by white nation-builders in Australia and South Africa' when it was published in New York in 1888.[40] Bryce was part of a small but influential group of Oxford acquaintances, who, having followed the American Civil War and travelled extensively throughout the English-speaking world, believed the 'colour question' was the greatest threat to both the British Empire and democratic rule. While based upon the 'problems' with the black population in the United States after slavery was abolished, these ideas were swiftly adopted to fit the more widespread threat of Asian migration and lent an academic rigour and intellectual justification to exclusion. Bryce, in particular, publicised the racially restrictive legislation being implemented in America at the time as a model for the settler colonies to adopt to keep Asians out. Chinese migrants were not simply potential competitors; if the self-governing colonies wanted to be democratic, it was best not to have mixed-race societies. The specifics about exactly who could successfully be assimilated and who was unacceptable still remained nebulous, but it provided a broad intellectual framework to justify an emphasis on Britishness or whiteness and Chinese exclusion, since part of the stereotype of the Chinese was their inability to assimilate.

Perhaps the most unifying factors for these diverse 'white' communities were the fears of disease and sexual miscegenation. When there was an outbreak of smallpox in Sydney in 1880, a mob attack on the Chinese community held responsible was narrowly averted. Nevertheless, an Anti-Chinese League was established as a result, which advocated the exclusion of all Asians from Australia.[41] Likewise, the size of the Chinese population and the fact that Chinese men tended to migrate without female counterparts led many commentators to worry about the racial implications for such populations. The Sydney *Bulletin* in 1901 complained of them 'because they intermarry with white women, and thereby lower the white type, and because they have already created the beginnings of a mongrel race, that has many of the vices of both its parents and few of the virtues of either'.[42] Increasingly, urban segregation was implemented to prevent the spread of either disease or miscegenation.

The increasing use of censuses only further provoked anti-Asian feeling. From the 1840s, statistics charted huge increases in Asian populations, causing panics; which only increased as these populations spread and Asian nations like Japan and China increasingly asserted their sovereignty. Unsurprisingly, newspapers often referred to Chinese and other Asian migrants as a ' "tide", "horde" or "invasion" '.[43] The statistics, however accurate they actually were, 'proved' an increase of Asians at a time when Canada's population lost more migrants than it gained between 1861 and 1901 and Australia, after the 1880s, had a stagnant population. In southern Africa, not only were whites outnumbered by native Africans, the two 'white' races were themselves divided; in Natal, Indians outnumbered all European migrants, and indigenous Africans outnumbered all other groups by a significant margin. As Samson has explained: 'The West had long been constructing the Chinese empire as corrupt and effete, concluding that increasing European domination in China was natural', but wars between China and Russia, the Boxer Rebellion in 1900 which targeted Europeans in China, and the censuses decreased the confidence in such a view, especially as social Darwinism spread.[44] Although Europeans were reluctant to accept such an idea, European settlers increasingly described Asian migration as an invasion by stealth, a view most famously expressed by the politician, Charles Pearson in his book *National Life and Character* (1893), which used censuses and other 'hard' evidence to 'prove' that white supremacy could not be assumed and that its greatest menace came from China.[45]

When coupled with fluctuating economies and mobile white populations, such beliefs led to expulsion and exclusion. The Chinese were only 1.75 per cent of the entire New Zealand population in 1870, but 6 per cent of the population in the gold-rich Otago. In an area where the number of white settlers was already almost equal to Maori residents, the Chinese population was seen to tip the balance away from the white populous. So despite the small numbers, the colony implemented a series of restrictions from 1881.[46] Even Tasmania and Western Australia, which had less than 1000 Chinese in each, passed legislation in 1886 and 1887, respectively, largely in solidarity with their neighbours.[47] These numerous stereotypes used to describe the Chinese emphasise the strength of the networks of migrants and ideas which connected settlers who increasingly described themselves as 'white', and this groupness was in turn encouraged by the growing sense of a shared 'yellow peril'. In reality, most of these places did broadly share similar migration flows, but as the networks of stereotyping grew, the issues became less about the statistics of migration and more about a shared ideology and identification.

Colonial cooperation

Given the growing sense of peril, increased cooperation to better implement exclusion was the next natural step, especially for those places linked under the British Crown, with similar legal and cultural systems. In 1879, the Inter-Colonial Trade Union Congress in Australia 'unanimously' condemned any further importation of Chinese labour and called for the introduction of a heavy poll tax on those already resident.[48] The following year, the first ever British Inter-Colonial Conference was held. Although the British government encouraged trade as the main point of discussion, how to deal with Chinese migration was as prominent a concern for the colonial representatives. A motion was even passed which warned 'that in the opinion of this Conference the grave consequences which must follow the influx of large numbers of Chinese call in a special manner for the concerted action of all the colonies, both in representations to the Imperial Government and in local legislation'.[49] The motion was carried unanimously.

Such pacts did not signal, of course, that the colonies agreed on what course of action to take but they were acknowledgements that some exclusionary measures were desirable and that a common effort was needed to implement exclusion effectively. Nor was this an indication of specifically nationalist sentiments. Most colonials still adhered strongly to their Britishness; indeed, Lord Carrington, the Governor of NSW explained his Brycian view to the CO, that 'if these Colonies *are to be an offshoot of Britain*, they must be kept clear of Chinese immigration'.[50] Maintaining their Britishness required exclusion.

The influence of these beliefs on political action is most evident during the federation of Australia in 1901. This federation reflected the decades of information networks which bound it to other 'white islands' in the Pacific, especially the United States. Prime Minister Edmund Barton quoted extensively from Pearson's *National Life* at the inauguration of Australia's first national parliament.[51] Reportedly, all the legislators carried copies with them throughout discussions. They even called the Australian federation the 'Commonwealth' of Australia, alluding to the influence of Bryce's *American Commonwealth* on their 'white Australia' policy. This widely appealed to the political elites in the colonies perhaps because 'if Asia was viewed as the centre of a coming world conflict, it followed that Australia was at the cutting edge of the struggle for racial supremacy rather than an insignificant spot on the remote periphery of the British Empire'.[52] This network very much depended both on the authority of Britain and the desire to 'provincialise' it, to make them all equals, perhaps even to make themselves the centre of a 'white' world, rather than at the margins of it.

Unfortunately, colonials increasingly came to feel that it was the British government itself which was preventing them from securing their white democratic potential. While the majority of white residents in the settler colonies were being won over to the notion of Chinese exclusion, there was widespread support in Britain for a legal tradition which, at least in theory, was racially blind. This was a tradition of national pride which British humanitarians obsessively sought to protect, along with the principles of laissez faire. Colonial policies when covered in the British press were often portrayed as products of democracies which allowed ill-informed and hysterically prejudiced working-class people to have political power. This of course reflected class issues within Britain, but it also showed the lack of understanding or interest in Britain about why the Chinese seemed so threatening to their colonial cousins. After all, until the twentieth century, Asians were simply not a physical reality in most Britons' lives. Most of the Chinese and Japanese in Britain were sailors; Indians were slightly more numerous, but were usually either from educated elites or limited to port cities.[53] Furthermore, most of the British coverage of Asians was detached from any notion of an 'Asian menace' which threatened Britain directly. Charles Dickens had introduced the image of the East End Chinese opium den in London in the 1870s, but Gilbert and Sullivan's comic opera *Mikado* was a more common view. G. M. Trevelyan expressed a common contemporary view when he wrote that there was far greater danger from the white European influx into Britain, or from French and German competition, than from any 'yellow peril.'[54] Periodically, events like the 1857 Indian Mutiny or the 1900 Boxer Rebellion created brief shared concerns over the 'Asian menace', but such fears quickly evaporated in Britain.

This division in perspective between the 'white' colonies and metropole became all the more important in the 1890s when growing numbers of 'free' Indians, Afghans, Pacific Islanders, and Japanese migrated. When Britain signed the 1894 treaty with Japan, London had included an article which gave the self-governing colonies the option of ratifying it. While Queensland later did for economic reasons, the other Australian colonies agreed at the 1896 meeting of Australian premiers, 'not just to refuse the treaty, but to legislate against Japanese migrants'.[55] The NSW legislature then passed an extension of their 1888 Chinese Restriction and Regulation Act to include *all* Asians and Africans in 1896. This measure was first proposed at the same Australian Inter-Colonial Conference 'as part of an Australia-wide plan to bring to logical completion the defence against coloured immigration' and in preparation for future Australian union.[56] New Zealand also tried to pass an Asiatic Restriction Bill in 1896, which would have kept out all Asians (although not without some strong objections that Indians from the Indian Army would be included).[57]

However, as both pieces of legislation referred so specifically to restricting immigration on the basis of race and made no exceptions for British subjects or allies, the governors of the colonies took the very rare step of reserving them for CO approval. The situation put the British government in an awkward position, as Chinese and Japanese officials and the India Office protested against the legislation; nor did the British public support such race-based law. Indeed, most British newspapers simply ignored the whole issue, reflecting the lack of public interest generally.

While the Colonial Office debated how best to resolve the situation, the colonial backwater of Natal came up with an ingenious, if controversial, compromise. Despite the substantial economic benefits to Natal, upon gaining responsible government in 1893, their first act was to cease government subsidies for the indenture scheme, although they stopped short of an outright halt due to economic concerns. While the traditional image of Natal's European population during this period is of an almost purely first-generation British population deeply loyal to Britain and spurred on in their patriotism by the recent Jameson Raid in 1895, such was the scale of public hysteria after several reports of plague arriving from India in 1896–97 that a small but passionate section of the population began to threaten cessation if Britain refused to allow restrictive legislation.[58]

The Natal Prime Minister, Harry Escombe, eager to prevent a full-scale revolt, copied almost word for word the US Immigration Act of 1891, which had restricted 'classes' of migrants such as the infirm, criminals, and paupers. He also lifted from the American legislation an education test first put before the US Congress in 1891, which called for a reading and writing test for all immigrants over 16 years of age in their native language, itself based on earlier literacy legislation introduced in the southern United States meant to decrease the number of voting African Americans. Escombe considered Indians too clever for the same sort of legislation, so he wrote that the test would be in 'any language of Europe'.[59] This would enable an immigration official to make the tests easier or harder depending on their own views of the desirability of the potential migrant, based seemingly on class and education criteria rather than skin colour.

In this legislation, the CO also recognised the perfect way to avoid conflict. Joseph Chamberlain, Secretary of State for the Colonies, wanted to bring about closer union between Britain and its settler colonies; estranging them over immigration matters could scupper vital plans like tariff reform. He saw the colonies of white settlement, not India, as the greatest asset the British Empire possessed. And Chamberlain, although theoretically against race-based legislation, realised that the only way Britain could maintain the support of the self-governing colonies was to let them have their own way in matters clearly deemed so important to them.[60]

When Chamberlain proposed an 1897 Colonial Conference to mark Queen Victoria's diamond jubilee, the colonies' lukewarm reactions reflected the mixed feelings colonials had about imperial government interference in their local affairs. George Reid, premier of New South Wales, 'attributed the strength of imperial sentiment in Australia to the fact that Australians had been left to themselves; there had been nothing to goad or change them'. Only Richard Seddon of New Zealand and Edward Braddon of Tasmania supported Chamberlain's desire for some sort of population-based parliament, which would regulate and spend an imperial tax. Wilfrid Laurier of Canada remained neutral to appease his French-Canadian supporters. Australia and the southern African colonies insisted that they could not consider closer union until they had been federated. It seemed as if nothing of substance would be negotiated; it was difficult enough to decide where to lay communication cables.[61] Yet at this conference, Chamberlain famously told the premiers:

> We quite sympathise with the determination of the white inhabitants of these colonies which are in comparatively close proximity to hundreds of millions of Asiatics that there should not be an influx of people alien in civilisation, alien in religion, alien in customs, whose influx moreover would most seriously interfere with the legitimate rights of the existing labour population... but we ask you also to bear in mind the traditions of the Empire, which make no distinction in favour of or against race or colour[.][62]

While not specifically vetoing the NSW legislation, he proposed the Natal language test as a more acceptable model. Although neither he nor the Natal legislature denied that it was specifically formulated to keep out Asians, he could still deny accusations that it racially targeted them.[63]

The legislation was not immediately appreciated by white colonials, however. Many were reluctant to adopt dishonest legislation, a reflection of what Robert Huttenback has described as 'a rather undefined dedication to fair play'.[64] Nor did all colonials believe in the general exclusion of Asian migrants. Companies continued to find the steady supply and cheapness of Asian labour attractive, and even in Natal, those who wished to halt indenture altogether were in the minority. Many simply wanted tighter controls or forced repatriation at the end of indenture contracts. Western and Northern Australia continued to import indentured labour for their plantations, due to the very small local populations and extreme climate. In the New Zealand legislature, a strong lobby argued that any immigration restriction bill unnecessarily pandered to the worst prejudices of the working class and wished to have at least Sikhs exempted from restrictions, given their perceived importance to imperial commerce and defence.[65] A growing number of educated Indians, Japanese, and Chinese settlers also began to protest through political petitions, the press, and in the courts.[66] Senior officials in Canada were particularly uncomfortable with the overtly racist and anti-liberal nature of much anti-Asiatic campaigning, blaming their own support for such legislation on 'democracy' – the public wanted it.[67]

However, with Chamberlain making clear that the Natal language test was the only acceptable legislation, the colonial parliaments gradually accepted it. By 1907, all of the 'white colonies' adopted some form of the Natal Language Test. Indeed, it was the first major legislative issue passed by the Commonwealth of Australia in 1901 and one of the first passed after South African union in 1910.[68] The 'White Australia' policies resulted in the Asian population throughout the Commonwealth of Australia dropping from 1.246 per cent of the total population in March 1901 before the first of the bills came into effect to only 0.82 per cent of the population in April 1911, although overall the populations had grown.[69]

Colonials clearly wished to be part of Britain's empire since they could have adopted less 'hypocritical' legislation if they had wished, simply following through on their threats of revolt. A few colonials would have welcomed such a move. But the majority in all these places, despite localised differences, wished to remain British and white. That was the very reason for the legislation. Colonials were legally emphasising the difference between being part of a British (or white) world and being a part of a multiracial British Empire. As such they were emphasising their place as new and better Britains, equal partners to Great Britain, whether she acknowledged it or not.[70] In 1902, perversely after the first real test of imperial unity during the South African War, these underlying tensions would focus firmly on the 'labour problem' in the Transvaal. After all that had happened, in 1902 the idea that a British administration would organise the migration of thousands of Asians to a 'white colony' seemed almost impossible to imagine.

2
The Transvaal Labour 'Problem' and the Chinese Solution

The labour problem

Why then were 63,695 Chinese labourers imported into the Transvaal when the other 'white' colonies were in the process of restricting Asian migration? To answer this question it is necessary to examine how the transnational issues played out in the previous chapter developed within the local context of the Transvaal just after Britain had won the South African War. The whole issue reflected southern Africa's position at the crossroads of African and settler colony policies and beliefs. On the one hand there was the 'civilising mission' to Africans and a need to promote colonial financial interests through encouraging African labour. On the other hand there was a concern about whether southern Africa could ever be a 'white colony', with such a large African population. 'It is nonsense to speak of South Africa as a white man's country at present in the same sense that Australia is. She has millions of aborigines, while the white number only as many hundred thousands', wrote one worried British South African.[1] Even if 'whites' were more united, they only made up a small fraction of the overall population, and yet British colonial policies specifically focused on turning the region into a 'white' colony.

These concerns became encapsulated in the phrases, the 'labour question' or 'native problem' in southern Africa. This terminology, the use of 'native' and 'labour' interchangeably to describe a 'problem' or 'question' of labour supply expressed the assumed connection between indigenous populations and unskilled labour since the beginnings of colonisation and slavery in the region, and was similar to the 'Chinese question' or 'problem' described in other colonies.[2] Both reflected colonial concerns about labour and racial hierarchies. How to actually encourage Africans to work was particularly problematic. If Africans did not want to work, what were colonials to do?

What to do with Africans, and a general labour shortage, were among the most pressing problems in the post-war reconstruction period.[3] Labour shortages had featured since the beginning of colonial conquest in the 1600s. The Dutch had imported slaves from eastern VOC possessions to carry out most work in the Cape, while locals were also enslaved or 'apprenticed'. By the twentieth century, 'the South African countryside' was 'a land... of large owner-occupied farms worked by a harshly exploited Black labour force'.[4] Labour was part of the basis for white supremacy: black men were civilised through the dignity of labour overseen by whites.[5] As the Rector of Johannesburg wrote in an article called 'The Native Problem in South Africa', given Africans were a permanent feature in the colony, 'the future welfare of the native is bound up with his ready acceptance of the yoke of labour'.[6]

This issue was of increasing concern to gold mines, the largest industry in southern Africa. While gold mining could be very profitable, the initial investments required were substantial and returns slow to materialise. Various financial scandals had reduced the confidence of foreign investors in 'kaffir shares' even before the war; most of the mines had been shut down or destroyed during the war,[7] and many companies were now considered overvalued.[8] Furthermore, the grade of ore to be found in the Rand was both difficult and expensive to mine. After the war, the mines had to dig deeper and they could not put up the price of gold to cover increased costs, as the global price of gold was fixed (in 1902–03 at 38 index points lower than before the war). While mine work resumed in March 1902, an excerpt from the Report of the Transvaal Treasury on the financial year 1902–03 revealed that 'only 29 mines... about one fourth of the industry as it stood before the War' were working.[9] The problem, according to the mines, was a lack of unskilled labour. This was not an actual shortage of people but a lack of men willing to work on the dangerous, badly paid mines.

Most efforts to tackle this problem did not focus on improving mine conditions but on securing labour outside of the Transvaal. Since the 1890s the general association of mines, the Chamber of Mines, coordinated their recruiting efforts through the Witwatersrand Native Labour Association (WNLA).[10] Unskilled workers recruited for the mines were almost entirely African, usually from neighbouring Mozambique.[11] The South African High Commissioner, Alfred Milner, had negotiated a new *modus vivendi* with the Portuguese in 1901 to ensure a steady supply and in 1902, when work on the mines recommenced, 66.3 per cent of all unskilled mine labour came from the east African Portuguese territories through WNLA recruitment.[12] This source of labour proved insufficient, however. The gold mines had stopped production for almost two years, and there was a great deal of competition from rebuilding and construction of new railways and roads (meant to assist British control and union), as well as from farmers and the diamond mines which often paid more and had the reputation for better working conditions (although they also suffered from labour shortages throughout 1902–03).[13] Postwar inflation and lowered wages meant that the same money did not go as far so mines were less attractive places to work.[14] Indigenous Transvaalers usually preferred agricultural work and only rarely worked on the mines for short periods of time before returning to their agrarian homes to farm for the remainder of the year. Other African recruitment efforts only supplied 1834 new labourers in 1903, turnover had increased and permission to expand recruiting within southern Africa looked unlikely.[15] Both the Cape and Southern Rhodesia had their own labour shortages and Natal forbade foreign recruitment of its labour; it imported Indian 'coolies' precisely because local Africans were so often unwilling to work on sugar plantations. The Foreign Office (FO) refused requests to recruit anywhere in Africa where they feared recruitment would be forced, such as Angola or the Belgian Congo.[16] By December 1903, the number of unskilled labourers on the mines was only 60 per cent of the pre-war level, which had, even then, been inadequate.[17]

Despite all of these problems, the colonial administration, headed by Alfred Milner and his 'kindergarten', had grand plans for the region.[18] These depended on increasing the British population so that it outnumbered Afrikaners and improving the economy so sufficient taxes could be collected and jobs made. Indeed the reconstruction period was named by contemporaries in direct reference to the social engineering projects in the United States after their Civil War.[19] For instance, railway rates were reduced so that living in the area would be less expensive for British settlers despite the resulting decrease in taxes.[20] Milner explained to the CO his plans:

> I attach the greatest importance of all to the increase in the British population... If, ten years hence, there are three men of British race to two of Dutch, the country will be safe and prosperous. If there are three Dutch to two of British, we shall have perpetual difficulty.[21]

His administration's focus on building a successful 'white', British colony was from the beginning perceived to be dependent on the gold mines. Their position as the largest employers and largest tax payers in the colony made them a natural focus for administrative efforts. As locals were unskilled in mining techniques, workers came from Europe, America, or Australia; a clear majority were from Cornwall.[22] They would travel to the Rand to work for a few years and then return home; permanently settling on the mines with a family was exceptional, due to the high cost of living. Milner's government wanted to encourage more permanent British settlers by assisting immigration and reducing living costs. Milner believed that the 'certainty of a vast and immediate expansion of mining and other enterprises after the war' would fund reconstruction and pay back the costs of war.[23] Increasing mine productivity could increase the taxes they paid to the government and the number of jobs available to British migrants.

Before the war the Transvaal (ZAR) government had largely resented the presence of the mines and the large foreign population which worked them, so it had done little to assist mine recruitment, giving preferential treatment to agricultural labour needs.[24] The new colonial administration could not be so sanguine, since reconstruction plans depended upon the mines increasing capacity. The most widely read newspaper in South Africa explained: 'a prolongation of the present financial crisis might mean a state of bankruptcy, and the Government could not afford to wait' to help secure unskilled labour, warning that 'depopulation, and perhaps starvation, would ensue'.[25] Thus it was that the solution to the post-war political and social unrest came to be seen by Milner's administration as dependent on fixing the 'labour problem'. The Native Affairs Department stopped short of recruiting African labour, but it did set up a pass department, appoint inspectors, and create a fingerprint database to better enable the mines to get and keep African labour on their mines.[26] Although some mines resisted, Milner formed an innovative system in which medical and welfare workers carried out inspections of the food, sleeping arrangements, and medical treatment available to miners. A committee of doctors was set up to find ways to reduce the death rate of mine workers, also at the request of Milner, and several suggestions from their report in June 1903 were implemented, meaning the death rate fell from 54 per 1000 in 1903 to 30 per 1000 by the end of 1904, still one of the highest in the world but a significant reduction in a very short period.[27] He also brought the mining area under the control of the Johannesburg municipality, making them accountable to local government and requiring them to pay rent to the Town Council for the first time, ensuring that at least some taxes were collected, and that these taxes could fund the greater oversight.

This represented a wholly new and unprecedented level of government interference in industry practices within the colony and also contradicted suggestions that Milner was merely a puppet for mining interests. Rather, he was happy to use government power to shape society.

In line with this, in 1903 Milner helped obtain a £30 million loan directly from the British government, backed by the Bank of England at an especially favourable interest rate of four per cent. The mining groups promised to pay it back in three annual instalments of £10 million, in recognition of their support for British rule in the colony and their own economic importance. Chamberlain arranged for a further 'loan of £35,000,000, bearing the guarantee of the Imperial Government, and secured on the general revenue of the Transvaal and Orange River Colonies' at three per cent interest per annum. Both were to be paid back in instalments starting in January 1904.[28]

All reconstruction plans depended on the mines, which in turn required enough unskilled labourers. Both the government and the mines perceived the 'problem' and the solution entirely through a series of racial assumptions. The British administration was motivated by their plans to cement the colony as 'British' while the entire notion that there was an unskilled labour shortage was based on assumptions about the racial characteristics of 'labour: white, black, and yellow', as many commentators at the time discussed. A government-commissioned report into the shortage of African labour decided

> that the scarcity of labour is due first and mainly to the fact that the African native tribes are, for the most part, primitive, pastoral or agricultural... The only pressing needs of a savage are those of food and sex, and the conditions of native life in Africa are such that these are, as a rule, easily supplied.[29]

A recruiter in Mozambique thought the problem was 'on account of the indolence of the native'; they were simply too lazy.[30] Milner wrote: 'South Africa, it becomes increasingly evident to me, has not got labour enough, even if the blacks were more industrious than they are.'[31] In addition to the stereotypical view that Africans were lazy and had few 'civilised' needs so required little money, commentators also blamed the war. A prominent British mine director, Percy Fitzpatrick, thought that the shortage of Africans was brought about by 'the position of affluence and comfort which the native at present enjoys owing to the excessive rates paid by the military and the bountiful harvests of the last two years', ignoring the actual food shortages and cattle diseases which were affecting Africans throughout southern Africa.[32] Even the official

Report by the Commissioner of Native Affairs complained in 1902 that 'the war had demoralised many of them [Africans] who had, in consequence of the familiarity shown to them by the British soldiery and of receiving inordinate high wages, become insolent and over-bearing'.[33] Furthermore, the war and growing mining community had decreased the number of Afrikaners with access to farm land: 'Anxiety about these widening fractures in the burgher community was reflected in populist agitation against foreign capitalists and black tenant farmers, who together were supposedly taking over the land and ousting the whites.'[34]

These views of Africans probably led to the biggest reason for poor recruitment in the post-war period. The Chamber decided in 1901 to slash African wages to 30 shillings per week, down from an 1899 average wage of 50s. The idea was that this would force Africans to work longer to secure the money to pay their taxes. 'The higher the rate paid to the native for his labour, the sooner has he secured his ideal' and would return home, explained a former Inspector of Mines for the ZAR government.[35] The mine owners assumed that Africans would not care that they were being paid less money for the same amount of work, because Africans did not think like that. Unsurprisingly with hindsight, this negatively affected recruiting and the subsequent scarcity of unskilled African labour drove wages back up by an average 36.9 per cent within a year. Consequently, wage expenses increased rather than declined and the shortage continued.[36]

Labour solutions

The obvious economic solution was to restructure the labour system so that it was not inhibited by race-based hiring policies. If the mines had really been motivated purely by economic considerations, this would have been the ideal solution. Instead, while there were isolated attempts to employ Africans in skilled or semi-skilled jobs, this was never widespread, and Africans were still paid less than equivalent white labour. This also led to widespread objections from local white unions worried about protecting their higher wages.[37]

There were more serious attempts to use unskilled white labour. The idea proved especially attractive because it could theoretically provide jobs for British settlers and decommissioned imperial soldiers, or even Afrikaner *bywoner* (those without land). Creating jobs for Britons had been a significant part of the justification for the South African War. An Australian miner on the Rand explained that 'white labour is cheaper and better. All Australians [believe] that...the work can be done as cheaply, and better, with all white men. White men would work together. They know how to work, and they know what is wanted'.[38] Their greater intelligence and racial solidarity would ensure a greater efficiency than African labourers were capable of achieving, so the theory went. At no point was it considered, by members of the public, the colonial administration, or the mines, to offer white unskilled labourers the low wages paid to Africans. Instead, it was widely assumed that a 'civilising' wage was essential for any white worker. 'The wage they required must provide not only for their own subsistence but also for their families subsistence and reproduction', something which was not factored into African wages.[39] Africans were not considered 'civilised' so did not need the higher wages.

Several experiments began almost as soon as the war was over, using decommissioned soldiers on railroads and gold mines; convict labour was even tried briefly by the government. On the mines, the largest experiment with white unskilled labour was spearheaded in 1902 by Corner House, since 1888 the offices of Eckstein & Co., the Johannesburg partners of the gold financiers, Wernher, Beit & Co., making them 'not merely the largest but overwhelmingly the richest mining group'. They were also the most closely allied with the colonial government and the strongest supporters of the war and of Milner's reconstruction plans.[40] Their scheme was headed by the engineer, Fredrick Creswell, the future co-founder and leader of the South African Labour Party.[41] His experiment took place on the Village Main Reef Mine, the most profitable mine on the entire Rand, but to justify the higher wages paid, they had to be three times as efficient as their 'native' counterparts. Unfortunately, recruiting figures were poor, turnover high and despite racial expectations, they were not efficient. Locals refused to do 'kaffir work' unless absolutely destitute and for very short periods.[42] Worse still for Creswell's experiment, his workers went on strike during September–October 1902 when he tried to increase their workload in order to boost efficiency. Labour unions encouraged the strike, fearing the experiment undermined their own position and pay as a white elite.[43] The project operated at a loss until it was finally cancelled in March 1903.

While it seems clear that racial assumptions had blurred economic sense, there has been intense debate ever since about whether white unskilled labour really was a viable solution to the 'labour problem'. Indeed, this aspect of the reconstruction period has been one of the most studied, especially because several mining magnates publicly declared their desire to limit the power of labour unions.[44] While labour unionisation and class prejudice were clearly concerns for the mines, this fails to take into account the deeply held racial assumptions of those involved with the scheme. Elaine Katz, in the most detailed analysis of the subject, has argued that 'the [white labour] scheme was unsuccessful principally because... the wage was too low to provide a single man, let alone a married man, sufficient income for subsistence'.[45] It needs to be emphasised that it was not that no man could survive on lower wages, however; it was assumed that *white* men could not. African workers were paid far less yet no one seriously suggested lowering the wages paid to unskilled white workers to the same level as their African counterparts. Those for and against unskilled white labour were fundamentally arguing about how best to promote the racial supremacy of whites. Creswell and his supporters thought white labour superior to black and championed a labour unionism based exclusively on whiteness. Mine experiments were also based upon a highly racialised set of assumptions, as was the cancellation of the experiment. The Chamber of Mines Annual Report for 1903 stated that the mines were finally unanimously 'opposed to the introduction of unskilled Europeans at such wages as would not admit of their existence as *civilised* members of the community'.[46] Lionel Phillips, a director at Corner House and originally a keen advocate of the scheme, even published a best-selling book called *The Labour Problem*, urging the 'paramount importance' of maintaining the race-based 'proper spheres' of labour division.[47] J. P. Fitzpatrick, another Corner House director and one of the strongest supporters of Creswell's scheme, did not envision it as a long-term solution but merely 'a stepping stone to better things' for British soldiers still in South Africa.[48] Milner was especially dismissive of using white unskilled labour, explaining, '[o]ur welfare depends upon increasing the quantity of our white population, but not at the expense of its quality. We do not want a white proletariat in this country'.[49] Even the existing white miners in the Transvaal largely viewed the scheme unfavourably, because of their own desire to protect their elite position economically and racially. As L. S. Amery later recollected, Creswell's 'claim was rejected by every other mine manager and engineer, and in the end their conclusion was reluctantly accepted' by the colonial government.[50]

While white unskilled labour proved unsuccessful, efforts were made to improve working conditions for African labourers and to develop the efficiency of the supply available. The mines, with the cooperation of missionaries and other moral crusaders, had passed Ordinance 32 of 1902, which prohibited the distillation or sale of spirits for 'native' consumption, arguing successfully that natives were particularly susceptible to 'vice'. This worked well in improving the efficiency of the workers they had but made the mines a less attractive place to work. Silicosis was killing off labourers at an alarming rate; the average life expectancy of Cornish miners on the Rand was only 36.4 years in 1904 so there was a high demand for new recruits.[51] There were also many attempts to introduce new equipment, such as tube mills and mechanical drills. These were all long-term solutions, however, and the technology was expensive and created new problems. The mechanical drills, for instance, made drilling faster but there was more wastage per ton mined and required specialist workers, so it was never fully utilised during this period.

Chinese labour considered

The idea that Chinese labour might be the solution to the 'labour problem' was first discussed at a Chamber of Mines meeting in 1898, reflecting similar debates about Chinese labour elsewhere. While Chinese labour was 'efficient and cheap...it would be inadvisable to introduce such labour' due to the controversy of Asian migration generally.[52] The stereotypes, prominent in other settler colonies, were well established, especially among the more than 5000 Australians on the Rand (10 per cent of the British Transvaal population) and through the newspapers. Such sentiments were so common that, during the South African War, the Afrikaner and Chinese races were often compared disparagingly in the English-language presses.[53] In Cape Town, 'Asiatics collectively were called "parasitic hordes"'.[54] Natal had begun implementing its language test. Any push to import Chinese labour to solve 'the labour problem' was thus bound to be controversial. With the unfriendly ZAR government in control at the time, the issue was shelved until after the war.

The first serious post-war push for such labour came in late 1902 from Hennen Jennings, an American mining engineer for Wernher, Beit & Co., who suggested that indentured Chinese men could work the mines and Africans would be freed up to be farm labourers.[55] While several leading mining figures such as Percy Fitzpatrick, chairman of the Chamber, and Harold Strange, chairman of WNLA, were against the idea,[56] it was agreed at the February 1903 annual meeting of the Chamber of Mines to explore their options further. They would intensify African and white recruitment efforts, while sending a delegation of mining experts to California, the Far East, British Columbia, and Malaya, to see if Chinese labour would really be suitable and available if wanted.[57] As the London offices of Werner, Beit & Co. explained:

Of course, you cannot import Chinese at the moment, because the town and the small trader are unquestionably dead against it, but the Chamber of Mines, as the responsible body for the mining industry, should all the same adopt a more decided attitude... the whole immediate future of South Africa, politically, socially and financially, *depends solely* on the mining industry.[58]

By then, some members of the Chamber had also completed a rough draft of a labour importation ordinance for public and internal consideration. The author of the ordinance was George Farrar, Chairman of the East Rand Proprietary Mines (ERPM) which controlled the largest single block of mining ground on the Rand (and employed the largest number of labourers), although the company was not as well off as the Corner House and Wernher, Beit & Co. partnership, and had personally received intense negative publicity, especially in France, as a result of the lack of dividends ERPM was paying, which possibly explains his eagerness.[59]

To assume that he was solely motivated by economics is, however, misleading. Farrar, along with Abe Bailey, James Percy Fitzpatrick, and Lionel Phillips, were some of the biggest names within the gold mining community in southern Africa who also engaged in politics. All were British, all were involved in the failed Jameson Raid in 1895 and all fought in the South African War. The political involvement of these men was atypical; both Fitzpatrick and Farrar received stern warnings from senior partners, investors, or financiers in London for their political involvement on more than one occasion and both had risked their lives to secure British rule in the Transvaal. Most other senior mining figures steered clear of such overt political involvement. Farrar, however, embraced the role of promoting British interests in the region. Milner had even appointed him to the Transvaal Legislative Council in 1903, and Fitzpatrick and he were the first mining magnates to be knighted in 1902 for their loyalty to the Crown. Farrar would go on to lead two major political parties in South Africa, both of which advocated closer union with Britain and stressed British racial superiority. It was his popularity and determination that ensured Chinese labour went from being an improbable solution to being seriously considered. Chinese labour was not the 'brainchild' of Milner, as one historian recently called it,[60] nor was it a conclusion reached unanimously by the gold mines in order to save expenses. While Farrar was not the only person advocating Chinese labour, he became the public face of the Chinese labour proposal. And again, while money was clearly a factor, his imperialistic world view cannot be dismissed. He wanted to find labour in whatever manner would make money and secure the colony for Britain. While the mine owners were still far from decided upon the matter, Farrar became the main champion of the proposal publicly.

He was assisted behind the scenes by Milner. In March 1903, Milner appointed men to an official Labour Commission to determine if sufficient African labour existed and how best to recruit it. At the same time, Milner was probably persuaded that labour importation was necessary. When the Transvaal government's accounts moved more than £500,000 into the red in mid-1903, he increasingly saw Asian labour importation as the solution to the economic problem (and thus the other problems) of southern Africa.[61] His favourable view of such importation was tempered by a desire that any such immigration would be small-scale and experimental in nature, not a permanent addition to South Africa's population. He shared many of the common prejudices against Asian migration so worked with Farrar throughout March and April to construct a series of restrictions so that 'the danger of the Asiatic spreading all over the country, settling here permanently, and adding to our already numerous racial difficulties, would be minimised'.[62] Asians were needed because, 'without the impetus he would give, I do not see how we are to have that great influx of British population...which is the ultimate salvation'.[63] The Chinese were needed because of various racial stereotypes: blacks were lazy, whites needed to maintain their dignity, and the Chinese were cheap and available and hard working. Restrictions were needed because of the widespread assumptions about racial conflict discussed in the previous chapter. And once Milner made up his mind, he acted upon it. John Buchan wrote years later 'when [Milner] had satisfied himself about a particular course...his mind seemed to lock down on it, and after that there was no going back. Doubts were done with, faced and resolved, and he moved with the confident freedom of a force of nature'.[64]

The Bloemfontein conference

Milner and Farrar eagerly sought to secure support within southern Africa from official representatives from each colony. Milner arranged an Inter-Colonial Customs Conference for March, modelled on inter-colonial conventions in Australia, primarily so the Transvaal, Cape, Natal, Orange River Colony (ORC) and Southern Rhodesia could discuss shared concerns over tariffs, railways, the labour shortage, and to pave the way for future union. Each colony sent representatives (from the elected parliaments in the Cape and Natal; appointed by Milner in the Transvaal, ORC, and Southern Rhodesia). As a Transvaal delegate, Farrar used the opportunity to submit a resolution in favour of Asian importation, while Milner apparently remained quiet about his support for the scheme and his involvement in writing the resolution during negotiations. It took 13 days of debate (the conference was only three weeks long) before a resolution was finally passed unanimously:

> The permanent settlement in South Africa of Asiatic Races would be injurious and should not be permitted; but that, if industrial development positively requires it, the introduction of unskilled Asiatic labourers, under a system of Government control providing for the indenturing of such labourers and their repatriation at the termination of their indentures, should be permissible.[65]

Partly, this was to ensure the support of Southern Rhodesian representatives eager to secure their own indenture scheme[66] and partly because Milner naively thought the British government might look more favourably on a scheme which used British subjects, such as Indians. Some mining magnates also preferred the idea of using Indians, although most favoured the reports of Chinese men's strength and discipline and worried Indians were British subjects who 'would only complicate the political situation on the question of [the] vote'.[67]

Once the resolution was passed, Milner cabled Chamberlain requesting a small number of Indians for work in government reconstruction projects in the Transvaal and for Rhodesian mines. However, when the request was passed on to the India Office, they turned it down immediately.[68] The India Office was already in a lengthy dispute with the Natal government over their treatment of Indian immigrants and had been with the former ZAR government over restrictions targeting British Indians; those restrictions remained in place under the Vereeniging peace treaty terms. Furthermore, objections were made to the repatriation requirement, as an illegal restriction on the movement of British subjects.[69] Chamberlain then vetoed the entire suggestion and emphasised that no scheme would gain approval until it was clearer that 'the essential features were acceptable to the general opinion of the white population', not just the 13 representatives, largely appointed by Milner, in attendance.[70] Moreover, the CO stated that Southern Rhodesia was not to be treated with the priority of a self-governing colony; instead, 'they must wait and see what was done in the Transvaal'.[71]

The labour reports

Despite these developments, efforts to import indentured labour were strengthened in the autumn of 1903 with the publication of two reports: the government's Labour Commission issued reports while the Chamber's report into Chinese labour was also published. The Transvaal Labour Commission had interviewed 92 witnesses between 21 July and 6 October 1903, after advertising 'throughout South Africa by post and by insertion in all prominent South African Newspapers' for anyone with information or viable solutions to the labour shortage to come forward.[72] When the findings were published on 19 November, the committee could not agree, however, so two reports were issued, ten signing the Majority Report and two the Minority Report. The Majority Report did not discuss Chinese labour, simply that African supplies were deficient and white labour could not fill the gap; furthermore, it decided the labour shortage was endemic to South Africa, not a result of the war.[73] The Majority Report became the official interpretation of events for decades to come when it claimed 'that the scarcity of labour is due first and mainly to the fact that the African native tribes are, for the most part, primitive, pastoral or agricultural... The only pressing needs of a savage are those of food and sex, and the conditions of native life in Africa are such that these are, as a rule, easily supplied'.[74] Furthermore, 'when a native gets an idea into his head it takes a long time to remove that impression', meaning that when Africans had heard that the mines were paying less, it took a long time to change that impression, the 'native' being naturally slow.[75] Since Africans could not be compelled to work ethically, labour would have to be found elsewhere.

Their views were not without controversy, however. All of its signatories were directly or indirectly connected to the mining industry and nine of the ten had already expressed themselves in favour of Chinese labour before the Commission results were published: A. Mackie Niven (Chairman), J. Donaldson, W. Leslie Daniels, G. H. Gogh, J. W. Philip, J. C. Brink, Samuel Evans, E. Perrow, C. F. B. Tainton, and George Farrar. Their report also focused on the benefits to 'the European investor' and the necessity for the mines to make more money for these investors, rather than what was in the best interests of the colony or its residents. The largely British panel were also clearly biased against Afrikaner testimony, especially criticising the suggestion from witnesses such as Louis Botha to use forced African labour.[76]

The Minority Report, meanwhile, was written by J. Quinn, a prominent baker, and Peter Whiteside, the Australian president of the Witwatersrand branch of the Trades and Labour Council (WTLC). Their report advocated white unskilled labour, accused the mines of exaggerating the labour shortage and refused to acknowledge the economic dependence of the colony on the mines.[77] They also argued that African labour was only scarce because African recruits were ill treated and because WNLA was poorly run.[78] Fundamentally, they thought the mines should 'be worked in the interests of the people of the Transvaal' and not just mine owners, something 'best secured...by the combined supply of white and African labour'.[79] Chinese labour was definitely not their chosen solution, as both had been actively campaigning against it since June.

Soon after, Farrar chaired a Chamber meeting to discuss the reports of the Labour Commission, which voted 'without dissent' to support the conclusion of the Majority Report and to request Chinese labour after a suitable amount of time had been allowed for public digestion of the report's information.[80] This was the first time that the mines unanimously agreed to support labour importation from China as their preferred solution to the 'labour problem'. This unanimity reflected a racialised economic view, which placed their faith in the myth of the Chinese 'working machine'[81] so beloved of industrialists around the world. While their own investigations into Chinese labour were not concluded until September 1903, favourable reports about the Chinese character and supply meant that at the June Chamber of Mines meeting, only one person present still opposed Chinese importation outright, but this was someone 'unconnected with any of the larger or smaller houses'.[82]

When H. Ross Skinner returned from the Far East in October 1903 and submitted his report to the Chamber of Mines, it was actually less favourable than many expected, highlighting both the positive and negative stereotypes associated with Chinese 'coolies' in California and the Far East. His tour had not involved any significant contact with Chinese people themselves but instead involved interviews with white managers or government officials with Chinese 'experience' and so stereotypes were largely untroubled by personal experience.[83] While still cautious, it stated that the Chinese 'are docile, law-abiding, and industrious people, and will carry out whatever contracts they enter into and perform the tasks assigned to them...The Chinese are, as a race, most easily led, if matters are fully explained to them'.[84] Skinner recommended 'that the immigrant is securely indentured and his repatriation rendered compulsory on the termination of such indenture'. He made clear that he thought the failure to do these things in the past had led to any friction between whites and Chinese. Repatriation was essential because of Chinese migrants' 'non-assimilation' into 'Western ideas and customs', 'the hoarding spirit inherent in the Chinese', their preference for trading 'with their own people', and their 'very low standard of living' compared to 'a white man, much less a white man who has to support a family'.[85] It was easy for the mines to gloss over the negatives and idealise them as a labour solution. The Chamber held a meeting to discuss Skinner's report on 6 October and unanimously approved his recommendations.[86]

After this, the Chamber was keen to press for Chinese labour as speedily as possible, and to cooperate with the government as much as possible to achieve this aim. The mines largely incorporated Skinner's advice into their ordinance draft published in October 1903, with the exception of racial and residential segregation, which was deemed too expensive and impractical.[87] After many such alterations, the mines were finally ready on 16 November 1903 to cable a draft ordinance to London. Continuing in an optimistic and conciliatory vein, the Chamber also paid to have the draft ordinance translated into Dutch and published throughout the Transvaal, so the general public could be reassured about restrictions.[88] As the Chamber told Milner, they were 'as anxious as anyone can possibly be to make immigration a success and are perfectly alive to the danger of introducing especially the trial cargoes on any but the most favourable conditions'.[89]

Fittingly, it was Farrar who introduced the Chinese importation legislation into the Legislative Council on 28 December. One by one, members of the Transvaal government testified to the labour shortage and need for Chinese labour: Sir Richard Solomon, Attorney-General, Sir Godfrey Lagden, Commissioner for Native Affairs, and Patrick Duncan, Colonial Treasurer. The debates last over 30 hours with some members speaking for more than three hours.[90] After the general idea of importation was finally agreed, the legislation had three readings in the legislature (a system modelled after Westminster) before gaining final approval, albeit with modifications. In the end, only four men voted against the proposal during the final reading on 30 December 1903, all staunch labour union supporters. The final draft of the Ordinance passed in February explained at great length the conditions of work, far more than previous indentured labour schemes had done.

Conclusion

While significant controversy over the scheme continued, the CO gave their consent on 16 January 1904.[91] Analysing the decision to import Chinese labourers makes it is clear that obsessions with the racial hierarchy played a significant part in the process. If the British colonial government had not been so keen to secure South Africa as a white colony, Chinese labour would almost certainly never have been allowed. The stereotype that Africans were lazy and slow-witted was equally crucial in the decision that there was insufficient African labour and that other alternatives would have to be considered. Similarly, if the mines had not accepted a racialised labour hierarchy and the stereotype of the Chinese as ideal working machines, Chinese labour would never have been requested. The idea that races had to be in competition with each other also ensured that Asiatic importation could be a threat to white supremacy and therefore that they would only be imported under tight restrictions. It also meant that white unskilled labour was never a serious option in South Africa, as it was already a firmly held belief that Africans and Europeans could not do the same work without threatening the racial balance. None of this, however, guaranteed public approval.

3
Greater Britain in South Africa: Colonial Nationalisms and Imperial Networks

When rumours began to circulate in 1902 that the mines were considering indentured labour importation, the recent Boxer Rebellion, South African War, and general controversy of Asian migration ensured global coverage. Nor was British opinion the only opinion which mattered. When the Chamber tentatively approached Joseph Chamberlain about Chinese labour importation in January 1903, he was not encouraging, explaining: 'The feeling at present all over South Africa is against such policy and, as long as this continues, it is not likely that the home Government would give its assent.' He insisted instead that the scheme would only gain approval if local public opinion could be proven to support it.[1] As the Transvaal legislature was entirely appointed by Milner, this alone would not suffice. How public opinion in the Transvaal was to be gauged without an elected government or a referendum, or indeed who the 'public' was, remained unclear. Were only whites, only British, only men, only Transvaalers to be consulted? The ambiguity of this and the existing controversies over Asian migration naturally attracted considerable attention in the Transvaal and beyond. Just like the wider issues of Asian migration, this reveals important debates about the nature of the empire, whiteness, and Britishness. The whole question of the role of the imperial government in determining the policies of individual colonies lay at the heart of the wider empire's concerns, as did the place of non-ethnic Britons within that empire. The reasons for this interest, the subsequent debates over the scheme, and the forums through which such debates occurred (petitions to the imperial and Transvaal governments, resolutions at public meetings, press articles, transnational organisations) are essential to understand contemporary attitudes towards imperial and South African federation. It also perfectly reflects pulls between local and global issues. There were shared sources and shared interests, but reactions were also distinctly localised. Nor can this be understood through the prism of metropole–colony relations as other analysis of the scheme have done. This was the closest the settler colonies and Britain ever came to federation, and debates over the scheme's implementation highlight the developing conflict between burgeoning ideas of federation and colonial nationalism. The pull of grand ideologies like 'white labourism', British 'liberalism', 'imperial federation', 'anti-capitalism', and 'free trade' were played out alongside unique, increasingly *national*, concerns in these colonies.

Farrar gave his first public speech on the subject in front of a few hundred mining employees on his own company's largest mine in Germiston on 31 March 1903, where he was likely to get a favourable reception. He stressed that importing Chinese labour would increase the number of jobs for 'white men, with their families'. He promised that he would 'never consent to any legislation on this question' unless the imported labourers were indentured, 'repatriated at the end of the contract', restricted only to unskilled labour on the mines and barred from any trading, farming, or other industry. Fifty-four jobs would be specifically banned to the imported labourers, including seven positions usually reserved for unskilled white workers.[2] This combination of restrictions had never been implemented before in the British Empire. It was not, however, an entirely new idea; some of the restrictions were based on existing legislation within the Transvaal, namely a bill passed in 1893 which declared that only whites could blast on mines[3] and one from 1886 which restricted 'Asiatic traders' to designated areas so as not to compete with white traders or spread disease.[4] These were laws which normally would never have been allowed by the imperial government but had been inherited from the ZAR. To the public, this meant that it was unclear whether such restrictions could be enforced, or indeed if they should be enforced by a British government.

It is unsurprising that many individuals and groups stepped forward to voice their opposition. As early as October 1902, when rumours were circulating about the proposal, shopkeepers set up the White League, an association seeking to combat Indian competition and 'to win the Transvaal for the white man and the white man alone'.[5] The group organised a Johannesburg meeting the day after Farrar's speech in order to object to Chinese importation. The main speaker was G. Hutchinson, who described his encounters with Chinese men on Australian mines in typically negative terms: 'Centuries of recession had debased them, had made them servile and cruel...They were industrious, they were frugal...they were diligent...[but] they were on a different footing to us.

To meet them in competition, we would have to cancel ten centuries of civilization'. Repatriation could never be enforced. This 'yellow horror' would live in 'their foetid dens' filled with 'the fumes of crude opium filling the air'. He then asked a question which would frequently resonate: 'Was it for this that Great Britain gave of her bravest and had accumulated a debt greater than all the debts of all the wars of last century? Was it for this that the Colonies of Australia and Canada had also sent the flower of their manhood to help us?' Another speaker argued that such a question 'did not only concern Johannesburg but the whole of South Africa'. Any decision in the Transvaal would affect their future partners in nationhood. Furthermore, unlike their neighbours, the Transvaal did not have 'representative institutions' to put their views forward; if democracy existed in the Transvaal, 'this question would not be so much as raised to-day, and they would, were it raised, give such an emphatic answer that there would be no trifling with the question'. This was a debate about preserving whiteness, health, and self-government, while also borrowing on the language of imperialism, anti-capitalism, and imperial federation. Despite the seemingly contradictory nature of these ideas, the crowd loved it. All but two of the estimated 5000 people in attendance voted for a resolution against 'the introduction of Asiatics'.[6]

The appeal to other colonies was not merely rhetorical. From the beginning it was clear that any proposal for or against Chinese labour would have to get approval from the imperial government, which could overturn any piece of colonial legislation. Because of this, most debates were directed not just at a local audience but at an imperial, especially a British, one. They had to convince the imperial government that popular opinion was on their side, either for or against the scheme. One Johannesburg anti-Chinese campaigner even wrote a pamphlet specifically for British working-class men, explaining the import of the issue both in terms of self-government and future federation:

> Each colony is self-governing, but in the strictly internal affairs of South Africa the people are not self-governing. They have no South African Government with which to govern. At present they are governed through, not by, the High Commissioner, who is not independent, but is subject to the control of the Secretary of State for the Colonies at Westminster, who again is subject to the control of the Imperial Parliament. Now, this is where you, the man in the street, come in. You elect that Parliament, and neither you nor they know much about South Africa.[7]

Union

One of the main reasons for widespread colonial interest was that it overlapped with debates about political union. After The South African War ended in May 1902, it was immediately clear that, at some point, Britain would seek to unite or federate the southern African colonies it now controlled. This had been discussed intermittently since the 1870s, especially in the Cape and within the CO.[8] Australia's own federation in 1901 was widely publicised in southern Africa and had renewed debates about future union. Chinese labour importation brought to the forefront concerns regarding the diverse legal and cultural approaches to race in each of the four southern African colonies. The Cape had a relatively liberal, theoretically race-blind system of voting, whereas the ZAR and ORC had a highly restrictive racialised system of governance, albeit one never adequately enforced, but which remained in place under British rule.[9] In order to secure military victory, the British had agreed in the Vereeniging peace treaty not to interfere with the 'native' policies of the Transvaal or ORC, but to leave it until self-government was granted to decide what changes should be made. This meant that specifically race-based legislation, which would not have been allowed by the imperial government normally, remained in place, such as a ban on non-whites voting.[10]

More widely, imperial federation was a growing area of consideration within the empire. Chamberlain was busy trying to promote his tariff reform ideas. Various organisations were also promoting the formation of an imperial parliament, with representatives from all the self-governing colonies (Australia, New Zealand, the Cape, Natal, Canada) and Britain,[11] a more formal embodiment of the colonial conferences already taking place. By 1902, the only major change which had occurred regarded defensive matters, with the South African War considered the first 'imperial' war. The British colonies in southern Africa as well as India, Australia, New Zealand, and Canada had all contributed personnel and money towards the war effort (occasionally given illegally to Afrikaners).[12]

No other issue could have been as controversial as Chinese labour importation in such a context. Asian migration was already widely perceived to threaten 'white' colonies, and it had long been associated with a struggle between the rights of the self-governing colonies and the imperial government.[13] Wider concerns about a 'yellow peril' ensured this story received special attention after peace was restored in May 1902. As one British MP remarked, the biggest problem facing the 'future of Empire is whether the British government should control and regulate the movements of people within it'.[14]

The battle for public opinion

Who the public was remained debatable, and it was equally unclear how public opinion was to be assessed. The idea that newspapers could be 'both a barometer and a guide of popular thinking' was widely believed at the time by figures such as Milner.[15] During the South African War, newspapers were used both to influence others to their way of thinking and to 'prove' public support for war. Men like Milner used the press during the war 'to control the sources of information and stifle the expression of opposing opinions which should normally be central to political or diplomatic exchanges'.[16] Much the same thing was attempted over Chinese labour, but on a transnational scale.

The Imperial South Africa Association (ISAA) was arguably the most active in promoting Chinese importation. Established in 1896 to promote the South African War effort, the ISAA 'aimed at transforming the English-speaking empire into a more consolidated and cohesive unit, a unit capable of holding its own in an age of intensifying military rivalry and cutthroat economic competition'.[17] Their funding came primarily from Tory backbenchers, South African mining magnates like Alfred Beit and imperialistic organisations like the South African League. Its successive chairmen were all Tory MPs, with strong personal loyalties to Milner. With the war won, it now turned its considerable resources towards promoting Chinese labour importation. Just as their efforts during the war had focused on 'Uitlander grievances' because it was assumed 'to be the safest way of putting the question of British supremacy in South Africa onto the agenda',[18] they focused on the need for Chinese labour to turn South Africa into a British colony. Andrew Thompson has shown that, by 1906, they had 12 agents and 34 speakers, 'visited 202 constituencies, addressed 505 meetings, and spoke on behalf of any candidate supporting the interests of labor in the Transvaal'. Dozens of pamphlets and approximately 'four million copies of each were circulated.'[19] Many of the speeches and ISAA pamphlets concentrated on explaining to a British audience why the mines could not use white unskilled labour. One man 'closely connected' with South Africa for 25 years, argued: 'In countries like South Africa or India, where a handful of whites are called upon to control by moral force hordes of savages it is surely in the highest degree undesirable – and dangerous – to place the white man in this position, or to put him on equal terms in labour or otherwise with the black man.'[20] A miner explained that 'Mr Cresswell, went kronk on it...everybody here knows that he is utterly wrong.'[21] Another simply stated: 'whites can't do kaffir work side by side with kaffirs'.[22]

The mines were the most organised in their attempts to influence public opinion and 'show' Britain that such support existed. In addition to using their newspapers, propaganda was organised through the Chamber of Mines itself. The Chamber formed a Labour Importation Agency (CMLIA) in mid-1903, which began organising petitions in favour of Chinese labour over the summer, gathered from meetings held on almost every gold mine on the Rand and from various towns like Belfast.[23] They also set up an office in London to lobby in October 1903. Their London secretary published and widely circulated a pamphlet arguing that 'the native had but few wants, and these were readily supplied with very slight effort' so Chinese labour would only 'supplement' their unskilled labour while 'natives' learnt to be more industrious.[24] Mining magnates also individually campaigned, through the pages of newspapers like the *Times*, and in a series of speeches and pamphlets.[25] Francis 'Drummond' Percy Chaplin, for instance, went on a tour of Britain in late 1903; he and Frederick Creswell followed each other across the country speaking at small town halls and to important political groups like the Reform Club in London.[26]

Opinion in the Transvaal did become less strident against the scheme, although a majority probably still objected. The threat of shutting the mines down does seem to have encouraged many miners that their jobs depended on securing Chinese indentured labour. The Transvaal Miners' Association (TMA) and the European Trade Union both declared their public support, while prominent labour leaders such Creswell, Outhwaite, and Hutchinson later admitted that Transvaal miners and traders were eventually won over by Farrar's campaign.[27] 'Only if Chinese labour comes they *say* there will be a boom', explained Emily Hobhouse disbelievingly,[28] while another writer thought that the Transvaal 'had been starved into acceptance of the Chinese'.[29] Miner explained to an Australian friend that there 'was a good deal of opposition to the introduction of Chinese here, but all but a few extremists are of opinion now that are the only solution' to the economic 'difficulty'.[30] Chinese labour might bring fresh dangers but, without it, there would be an economic disaster in the region. 'Our only chance of recovery...seems to depend on the alacrity with which Chinese labour can be introduced.'[31] The Johannesburg Chamber of Commerce eventually abandoned their January 1903 opposition in December because merchants were suffering under the depression and the situation had simply grown too urgent; a resolution supporting Chinese labour 'was carried without dissent'.[32] A general meeting of the Association of Mine Managers of the Witwatersrand passed similar resolutions in support of the importation.[33] The Johannesburg Stock Exchange passed one on 8 December, pushing for no delay to Asiatic labour importation, under proper restrictions.[34] The sense of urgency increased in early 1904 when the Russo-Japanese War began, as it was assumed no recruitment could happen during fighting.[35] Even the Anglican and nonconformist churches in South Africa decided to support the scheme, as it would be a chance to convert the 'heathen Chinee',[36] although their British counterparts largely opposed it as a form of slavery.[37] They even persuaded several South African and British nonconformist ministers to denounce accusations that the scheme was slavery.[38]

Labour unions & anti-Asian leagues

Those against the scheme were less well-organised but made concerted attempts to counter pro-Chinese publicity. The first widespread transcolonial campaigns against Chinese labour were organised by labour unions and their associated journals. The Amalgamated Society of Engineers' (ASE) journal, edited by an Australian, was one of the first to spread concern about the Transvaal proposals abroad; the organisation in Britain and Australia voted to support their Transvaal branch and Britain's ASE banned any Chinese from entering the society, even though none wanted to. Chinese members were already banned in Australia.[39] The *Transvaal Leader* sympathetically noted that anti-importation 'is strongly supported by the working classes', a view it shared, as did the Australian-edited *South African Trade Journal*.[40] Even the Johannesburg *Star* initially declared that it and the public were against Asian indentured labour.[41] Indeed, the editor of the *Star*, William Monypenny, chose to resign because he felt his own anti-Chinese feelings were in opposition to Corner House's position, which owned the largest shares in the newspaper.[42]

This was followed by a high-profile anti-Chinese campaign by the Trades and Labour Council (TLC) branches throughout southern Africa. Peter Whiteside wrote successfully to the Cape Town Trades and Labour Council for joint support against Chinese importation.[43] The WTLC and the Pretoria TLC also passed resolutions against the scheme, as did Australian branches of the TLC.[44] The racially mixed Political

Labour League, the first Cape labour party,[45] and a local Working Men's Union, set up and run by Isaac Percell, a local merchant,[46] were also active campaigners against the scheme. In July, an umbrella organisation, the South African Trade Union, unanimously voted that Asiatic labour 'was not in the best interest of South Africa'. The British Trade Union Congress also wrote and published pamphlets and articles, gave speeches and signed petitions against the proposal, with officials travelling frequently between Britain and South Africa to campaign for white miners' rights.[47] The annual conference of the Miner's Federation of Great Britain passed similar resolutions and declared in 1905 that they would not vote for a 'candidate who was in favour of Chinese slavery'.[48] At the meeting of the British Trades Union Congress (TUC) in 1904, their Parliamentary Committee also voted that all TUC members should 'cooperate in making a great national and effective protest against this return to slavery and insulting injustice of capitalism'.[49] The Johannesburg branch of the ASE voted against importation at their annual general meeting in September. Thomas Kneebone, organising secretary, thought the restrictions impossible to enforce: 'Unskilled...that is what they ask, but how long will they consider certain work as skilled is only as long as it takes John Chinaman to learn it...John Chinaman is clever and the best imitator born.'[50] The Transvaal Miners' Association (TMA) adopted a similar line at a specially held conference in October.[51] The Trades Union Congress also voted against the scheme at their annual 1903 conference 'because it was a question vitally affecting the [white] workers of this country'.[52] A very few members expressed concern for African workers, but most objections focused on preserving the elite position of skilled white miners against a capitalist-Asian onslaught. As one miner sarcastically noted, 'the war was, in a certain sense a miners' war – that was to say, it had been undertaken in order that justice might be done to the British miners of the Transvaal', but 'justice to the British miners of the Transvaal' required that 'British miners shall be kept out the Transvaal altogether'.[53]

It was easy for an imperial 'white labourism' to flourish in this environment. Published objections focused in particular on the unity of 'our own race' and that the issue was not a local matter at all but an imperial one. A typical objector asked: 'is the rest of the Empire, which shared with the Mother Country in the sacrifices of the war, to have no voice in a controversy which, *prima facie,* affects the objects for which the war was fought?'[54] If the imperial government sanctioned the importation, agreed upon by an unelected Legislative Council, it could eliminate the Transvaal both as a prospective workplace and a part of the 'white' colonies.[55]

Most of the Transvaal's labour unionists were from Australia, and this proved to be an important connection. Ideas of a global (sometimes British) 'white labourism' were put into action on a large scale for the first time.[56] Both the Australian and New Zealand governments were dependent on Labour Party support at the time, so the outcry from labour within these colonies immediately made them national issues. There had been some imperial industrial action previously, especially linking Britain and Australia, but this made an individual colony's labour situation the focus of transnational public attention really for the first time. In New Zealand, such public protests were even frequently organised by the prime minister, Richard Seddon, on the grounds that New Zealanders had 'laid down their lives' for British freedom in southern Africa, only to have 'hordes of Asiatics...sent into South Africa, to work the mines there, to take the place of the men of our own race, of our own blood'. The issue had been made an imperial one by their sacrifice, and so New Zealanders had to exercise their rights.[57] The Melbourne and Adelaide branches of the TLC thought it 'a menace to the British race and a disgrace to the flag under which we live'. The Melbourne Trades Half Council 'recommends all inter-State labor bodies to take immediate action, so that a united protest can be made'.[58] Whiteness was emphasised as creating a particular bond, giving them a right to interfere in Transvaal affairs. This embodied 'white labourism' and imperial federation in practice.

Not all anti-Chinese protests were made through labour networks, of course. Many highly vocal groups were formed specifically to fight against Asian labour importation; in the Transvaal alone, locals formed a White League, the National Democratic Federation, and the African Labour League. The latter was founded by independent businessmen, shop clerks, artisans, and trade unionists on the Rand in June 1903 to oppose the Chamber of Mines labour policies, especially Chinese labour. Most members were either recent immigrants who favoured white unskilled labour or long-time settlers who felt snubbed by the new administration, with lucrative jobs going to young imports from Britain, not 'old hands' like themselves.[59] On 30 June, the African Labour League held a meeting to create their manifesto, stating their belief that South Africa was 'eminently fitted to become the home of a great white nation'. They roundly condemned a biased press for overstating the need for labour and the economic difficulties. While drawing on the international imagery of white labourism, they also insisted that the

Transvaal should be granted its own government to decide for itself.[60] A few days later they petitioned the CO explaining their position.[61] They and other organisations like them focused on three particular aspects of the scheme, borrowing heavily from the language and experiences of white labour organisations in other settler colonies and the United States. First, they believed that the government ought to intervene to protect white labour, whereas they thought the government was intervening to protect mining interests, at the expense of their own political rights. Second, they reiterated all the traditional negative stereotypes about Chinese immigrants. Fundamentally, they insisted that such an important matter should be left in the hands of self-governing colonies. If the Transvaal was actually self-governing like the other settler colonies, the importation proposal would never be approved.

The 1904 Cape election

Outside the Rand, opinion was almost entirely against Chinese importation. At a company picnic, the Cape Railway Goods Department Commissioner gave a speech to his mostly British railways workers against 'the great yellow cloud'. He even expressed the rather radical view that while never previously considering woman's suffrage before, 'he wished they had the franchise for women to-day, for he felt sure that the women would take heed of the future and look to the interests of their children and children's children, and keep the Chinaman out of South Africa'.[62]

Cape residents had particular reason to be concerned about this issue. Since Britons first settled in the Cape, it had been the colony of primary importance in southern Africa, and with a race-blind electorate. However, with their increased economic dependence on the wealth of the gold mines, power was shifting away from them and there was concern that political union would mean the swamping of their interests with those of the gold mines.[63] If the Transvaal was politically dominated by the mines, then so could the future South African state. And for non-white residents, this could mean an expansion of the Transvaal's race-based restrictions. Such worries had been heightened not just by the war but also when Milner attempted to suspend the Cape constitution after the war. Milner had wanted to prevent Afrikaner 'tyranny'[64] in the case of an election before the political landscape had sufficiently calmed down. It was, to many, a direct act of government interference and the CO refused to approve it.[65] When the colonial government then collapsed in September 1903, an election was called.

There were two main parties in the Cape at this time: the Bond was generally a rural, Afrikaner party, while the Progressives were strongly pro-British and urban. As a result of racial antagonism of the South African War, John X. Merriman, a prominent liberal MP, headed a new South African Party (SAP). Merriman's ability to speak Dutch and his marriage to an Afrikaner secured the support of most of the Afrikaner population. While fundamental differences existed between the SAP and Bond, especially regarding the non-white population, they shared a vision of a united South Africa, where Dutch and English-speaking people were equal to one another, as the name reflected. In closely disputed areas Bond members were put forward as SAP candidates and in no constituency did they run against each other.[66] Chinese labour represented a perfect opportunity for the party to publicise and gain support.[67] The entire issue of Chinese labour was really about 'whether the development of South Africa is to proceed on the lines of one of the great self governing colonies or whether it is to become a dependency'.[68] These ideas were spread both through the electoral campaign and in the pages of the *South African News*, edited by another Oxford graduate, liberal, and SAP member, Professor H. E. S. Fremantle, the first history professor in South Africa. Fremantle too believed that it was 'unjustifiable to leave to a non-representative Government the decision of such a question as the present one'. Future federation gave all parts of South Africa the right to have a say in the matter, not just the Transvaal.[69] Miners were 'purely temporary inhabitants', whereas Cape residents were more likely to consider the good of South Africa as a whole.[70] Olive Schreiner's husband and fellow campaigner, Samuel Cronwright-Schreiner, thought 'capitalism' was the colony's true enemy and even suggested 'if Empire means control of the mining plutocrat, either directly or through Downing Street, then I've had enough of Empire'.[71] During one widely publicised speech in Cape Town, Merriman reiterated the foreignness of the supporters of Chinese labour: 'from where has the demand for the introduction of the people come? Not from the people of the country, not from those who make their homes in South Africa, but from financiers and shareholders who live far away from this land, and upon whom the effects – the detrimental effects – of these people will never come'.[72]

Such objections extended beyond the minority liberal British Cape population as well, reflecting the emphasis on southern African unity. F. S. Malan, editor of *Ons Land* (the most widely read Dutch-language newspaper in the Cape) emphasised to a Cape Town South African Liberal Association meeting the inevitability of being part of the empire and recommended instead racial conciliation. As one biographer has explained: 'Malan's dream was the development of an autonomous *national* character for the Cape Colony in which Afrikaners/Dutch and English-speakers could become Afrikaners in a South African nation.'[73] Otherwise, race would divide 'the permanent population of South Africa' and would allow the 'real enemy at the gate', capitalism, to 'threaten the very life of our civilisation' through the introduction of Chinese labourers.[74] Malan explained in an editorial: 'Our first object is to save South Africa from the ruin which it goes to meet if the foreign mine owners get their wish.'[75] Despite some Afrikaner misgivings about embracing British or African campaigners, most agreed with him, partly because suspicions of capitalism had long featured widely within local Calvinist church teachings. Dutch Calvinism 'condemned monopolies and other combinations as conspiracies against the consumer', and as such, the mining industry was often portrayed as inherently evil.[76] The great Bond leader, Jan H. Hofmeyr even came out of retirement to campaign actively against Chinese labour, citing the perils of letting capitalism gain too much power.[77]

Not everyone embraced this anti-Chinese effort, of course. Their chief opponents in the Cape were also their political opposition, the Progressives. The Progressives were now led by Dr. L. S. Jameson, notorious for the 1895 raid named after him and strongly associated with the mining industry. His party mainly consisted of British residents in the Cape; the very choice of Jameson as their party leader was seen as a show of patriotic loyalty.[78] Most were businessmen living in port cities like Cape Town or Port Elizabeth, particularly attracted to anti-Chinese stereotypes but unmoved by anti-capitalist rhetoric. Consequently, Progressive supporters' arguments focused on imperial 'loyalism'.[79] Milner said the colony wanted it; doubting him, claiming he was led by the mines, was fundamentally unpatriotic. The mayor of Cape Town, for instance, refused to chair a debate on Chinese labour because it would implicitly be a criticism of Milner's governance of the Transvaal.[80] The flamboyant mine magnate, Abe Bailey, advocated repeatedly in the Cape Assembly that, as the issue was currently before the imperial parliament, any Cape measures would be interfering and critical.[81] Furthermore, the *Cape Times* warned that if Chinese labour was not approved, 'the English skilled working men already in the country will probably never have a chance of getting to work at all'.[82] The Chinese were necessary to make the area British, while they alleged the campaign against Chinese labour was organised by Afrikaners who wanted to prevent South Africa becoming truly British. The Cape Governor, Sir Walter Hely-Hutchinson, warned that they 'hope[d] that the discontent may lead to a demand from the British section of the population for Responsible Government before the Colony has reached such a state of development as would render it safe...for the maintenance of British Supremacy'.[83] When Merriman launched an anti-Chinese petitioning campaign in November 1903, the *Cape Times* argued that while the public was free to express any concerns they might have, the Boer and pro-Boer press were duplicitous in their objections. Dutch farmers would 'be very glad' to get the Chinese, due to the shortage of labour, but would never admit it until after the election.[84]

By emphasising his party's argument that they supported real, albeit British, South Africans, not just the mines, Jameson had some success in early 1904 getting positive resolutions passed at town meetings. Port Elizabeth passed a petition stating that if it was allowed, the Cape must 'take such steps as will effectively safeguard the interests of this Colony' but would otherwise not interfere.[85] Aliwal North held one on 30 January, attended by Jameson, which passed 'by an overwhelming majority' a resolution along similar lines.[86]

Most towns took a different stance. The towns of Rhodes, Woodstock, Bedford, and Robertson all passed unanimous resolutions against any Chinese importation[87] while the tiny town of Tulbagh went so far as to 'resolve to use all possible means to oppose such importation'.[88] Middleburg passed a similar town resolution, although they specified 'every lawful means'.[89] Even 'an overwhelming majority' in the diamond mining town of Kimberley passed a petition against, drawn up after 'one of the largest Public meetings ever held in Kimberley'.[90] An amendment that the Cape had no right to interfere in Transvaal affairs failed.[91] These were largely British towns, but the public were clearly against the scheme regardless.

Even communities with few or no British residents in the Cape chose to send resolutions to the imperial government, in the hopes of influencing their policy towards Chinese labour importation. Cradock passed a resolution at a Dutch and English-language meeting, requesting responsible government in the Transvaal first.[92] In Cradock, there were Chinese men but they were businessmen while the 'hundreds of thousands' of indentured labourers would be 'of low caste' and 'beasts of burden' who would undermine local (white and black) workers and bring disease with them. There was also a 'native' contingent, all of whom supported the motion against the Chinese.[93] The Midden-Zwartland branch of the Bond unanimously adopted resolutions to support Afrikaner opposition to the scheme in the Transvaal itself.[94] The largely Afrikaner town of Humansdorp sent another unanimous petition complaining that the Chinese 'horde' would 'delude the country'.[95]

Unlike other South African colonies, voting was open to any man with £75 a year in property who could pass a literacy test, although communal tenure of property and all land in the Glen Grey district were disqualified. This still meant African, Indian, and Coloured men made up roughly 20 per cent of the Cape electorate in 1904 so, unlike in other southern African colonies, their opinions mattered.[96] While traditionally the Progressives were associated with a more liberal non-white policy than the Bond, the creation of the SAP and the Progressive's often aggressively patriotic dialogue confused this loyalty. Merriman made a point of including in his speeches information about how such a scheme would affect the indigenous population.[97] The *South African News* was keen to further his point: 'the natives will have no outlet for their labour and no material motive for progress... Instead of being one of the chief assets in the wealth of our young *nation* they will be its greatest danger, as any race would be, if it were then deprived of its natural function'.[98] When the SAP sent campaigners around the colony, they usually paid for Coloured or indigenous speakers to also canvass the country, a highly unusual activity even in the Cape.[99] Throughout 'Greater Britain' only the Cape allowed a regular forum for their opinions to be heard. They were, after all, 'native' South Africans in a way Asians or (Jewish) mining magnates could never be.

Perhaps the most high-profile African in southern Africa at the time, John Tengo Jabavu, a Cape Coloured teacher, Methodist lay-preacher and editor of the first African-run newspaper in South Africa, *Imvo Zabantsundu*,[100] was particularly active. He wrote and published a petition in English and 'Kaffir', wrote 'articles directing attention' to the campaign in his and others' newspapers and gave speeches on various 'native' locations around the colony.[101] At one of the largest meetings of non-whites ever at Fort Beaufort, he 'spoke of the importance of a movement against the Chinese' and gained unanimous support from the audience.[102] Even Dr. Abdulla Abdurahman, who had previously supported the Progressives and often been in conflict with Jabavu, campaigned on the issue. He felt disillusioned by both the Vereeniging treaty's failure to guarantee the rights of non-whites in the Transvaal and the proposed Chinese importation, believing it was meant to disenfranchise non-whites altogether. He subsequently switched his support to the SAP, although he remained an independent when he successfully ran as a Cape Town counsellor.[103] His support was particularly useful because he helped found the Political Labour League and also worked with a local Working Men's Union, which was unusually at the time mixed-race.[104] His efforts helped ensure that the SAP was able to coordinate not just with the Bond but also with growing labour networks throughout the Cape.

Other less famous Africans were also prominent in organising petitions and resolutions around the Cape, such as John Alf Sishuba, who arranged a large meeting in the Hackney, Oxkraal division of Queenstown, which passed a unanimous resolution against Chinese labour 'as tending to do away with the existing rights of labour and to stop our progress in civilisation and industry'. Sishuba and the Rev. Jonathan S. Mazwi jointly organised a similar meeting in Eardley, Queenstown, while Mazwi further organised unanimous resolutions in Kamastone and Oxkraal. The Oxkraal resolution was addressed to the King himself: 'Regarded as they do the interests of the whole of Africa as one and indivisible', they thought Asian immigration 'would assuredly tend to frustrate the efforts that are being made, not without success, to civilise and elevate the aboriginal natives by means of honest industry.' Furthermore, 'such an importation would decide whether South Africa is in future to constitute one of those great free communities under the British flag...or whether it is to be ranked as a mere plantation worked in the interest and for the benefit of foreign holders.'[105] These direct appeals to the King demonstrate how widely the issue in the Cape was viewed, both as something which concerned the South African *nation* and something which was inescapably imperial. While the self-governing colonies should have a say, it was ultimately the responsibility of the King to ensure that the empire lived up to British moral ideals – they were reminding the British government that fair treatment towards non-whites was supposed to be a fundamental part of this. This was the promotion of a *British* colony and Britain's 'civilising mission', portrayed as incompatible with foreign economic exploitation.

In light of public opinion, the Cape House of Assembly passed an almost unanimous resolution stating that Natal's situation as an 'Asian colony' might spread if further importation was allowed.[106] So strong were the anti-Asian feelings, by August 1903, the Cape Prime Minister could write to the CO that 'the Members of [the] House of Legislature are unanimous in their opposition' to the scheme and warning of possible revolt if the Cape was not protected through legislation from an 'Asian invasion' at the least.[107] Overall, electioneering was increasingly split between the SAP's argument that the Cape had a right (and duty) to actively campaign against Chinese labour and capitalist exploitation and a Progressive proposal to allow the Transvaal to decide for themselves while implementing local legislation to protect the Cape from 'Asian invasion'.[108] Despite the Progressives' arguments, the majority of the Cape population remained uneasy with the idea of thousands of Chinese being imported so close to their own borders, when border control was almost non-existent.

While ultimately Jameson's Progressives won the February 1904 election with a majority of seven, this probably had more to do with boundary charges and the large number of voters disenfranchised for fighting against the British in the war. In real terms, there was actually a shift away from the Progressives.[109] The last action of the Gordon Sprigg parliament was to cable that the labour shortage 'has not yet been proved to exist...the natives throughout Colony entertain strongest objection' and Ministers 'are convinced that great majority of white inhabitants of Colony are equally opposed to the proposed step'.[110] Likewise, the first act of the new parliament was to pass a motion 'that this House...reasserts its strong opposition to any such importation as prejudicial to the interests of all classes of the people of South Africa'.[111] Jameson also forwarded a petition with 15,244 signatures against the scheme.[112] Consequently, despite the election results, the only self-governing colony to have an election during the debate over Chinese labour was predominantly against the proposal.

Other voices

While non-white views only received serious attention in the Cape, many Africans widely believed that Chinese labour importation was meant to undermine their political rights and economic position in southern Africa. After all, the post-war labour shortage had actually led to a small increase in wages paid to them and gave them more choice over their terms and conditions. Importing other unskilled labour would further subordinate Africans in their inferior position in South African society. The notoriously poor treatment of Africans on the mines, and the mine owners' attempts to lower wages after the war made many Africans deeply suspicious of the plans. One chief, Sekhoma, denied that any labour shortage existed at all, believing instead that 'the mine owners mean to do away with the natives' altogether.[113] Chief Linchwe in the Cape said: 'They want Chinamen and not my people, who, they say, are too expensive.'[114] Just as Britons argued that the capitalists wanted to destroy white labour through the importation of the Chinese, many non-whites suspected this was an attempt to undermine their own fragile political and economic positions.[115] The Beaconsfield branch of the

South African Political Organization (SAPO) in Griqualand West sent a letter to the King, trusting that 'His Majesty will hearken unto the voice of his millions of loyal coloured subjects, who see in the introduction of Chinamen the greatest evils, for the peace, welfare and elevation of the coloured race.'[116] The Cape governor also forwarded a petition 'from representative gathering of natives in the Glen Grey district of Cape Colony imploring British government to refuse its sanction to immigration of Chinese into South Africa' as it would degrade the population.[117]

Chinese and Indians in southern Africa also expressed concern. A free Chinese man in Cradock, W. Manshon, wrote: 'my protest, together with my brother Chinamen here, against the importation of the lower class of Chinese labourers into South Africa' was because 'it would be detrimental to our interests as business men, and reflect [negatively] upon the present social positions we hold'.[118] M. K. Gandhi, working as a lawyer in southern Africa since 1893, used his newspaper, *Indian Opinion*, to express similar concerns. The newspaper was keen to stress their belief 'that the white race in South Africa should be the predominating race' while trying to emphasise a more class-based hierarchy, which would place educated Indians and Chinese above other groups.[119] They declared their opposition 'to Asiatic trading or to unrestricted Asiatic immigration',[120] and Gandhi even supported the Cape's anti-Asiatic immigration legislation provided 'the restrictions should apply only to non-British Asiatics'.[121] Repeatedly, the *Indian Opinion* explained that legislation which distinguished 'between British subjects and non-British subjects' or 'between the civilised and the uncivilised' people 'would be a natural division', but creating legislation which simply divided the population into 'white people and the coloured people' was 'most unnatural'. The restricted immigration of Chinese indentured labourers would foster colonial tendencies to racially divide the population in this way. When the scheme looked certain to go ahead in mid-1904, Gandhi switched to offering 'respect' for the decision but 'wished to ensure restrictions would not be applied to British Indians'.[122]

While such protests usually bowed to British moral and cultural superiority, they did not necessarily help the anti-Chinese cause or the political position of non-whites in southern Africa. The two primary canvassers for petition signatures in Grahamstown, organised by a local Anglican minister, were Coloured, but this apparently affected their ability to attract white signatures.[123] The episode also highlights the embedded prejudices against non-whites within the colonial administration. *Indian Opinion* articles were not passed on to the CO; their views were widely ignored. Likewise, the Cape was unique in having widespread anti-Chinese petitions from non-whites, which the Governor always passed on to the Imperial Government. Elsewhere, only British or Afrikaner opinions were forwarded to the British government for their consideration, when trying to gauge 'public opinion' concerning Chinese labour. Pointedly, one British colonial complained, 'it is an unedifying spectacle during elections to see the representatives of the two white races metaphorically wearing away the buttons of their waistcoats in crawling for the native vote'.[124] The *Cape Times* complained about the unrest such tactics might cause among all natives and coloureds.[125] In a secret report in 1904, the Cape government noted, 'a marked aggressive tone may be remarked in the Bond Press, together with much calculated to please the natives'.[126] Milner used the native vote to dismiss any Cape opposition altogether: 'The Bond is seeking to make up for the votes lost to it through disfranchisement by a bid for the native vote...The Progressives on their side cannot afford entirely to lose the native vote, so both parties are competing with one another in protesting against Asiatic labour.'[127] It was widely considered dangerous to direct political campaigns towards the non-white population, as the non-white population might then actively demonstrate discontent through political campaigning. Indeed, to an extent this is what happened, as the campaigning and petitioning strategies utilised were later adapted by many of the same people when the African Political Organisation (APO) formed a few years later.[128]

As a consequence of colonial and imperial government assumptions that 'the public' was the voting population, that is, white male Transvaalers, the government could not ignore Afrikaner views as it did non-whites views, however. Initially, most Afrikaners were less politically involved, partly because the government was suspicious of any large gathering of Afrikaners and partly because many were preoccupied with post-war reconstruction. Milner had asked a few 'bittereinder' generals – Louis Botha, Jan Smuts, and Koos de la Rey – to be part of the Transvaal's legislature when he appointed it in 1903, but they all refused.[129] Because of wartime hostilities, they did not trust Milner, nor did he trust them.[130] Instead, Milner appointed relatively unknown Afrikaners to the Legislature. J. Z. De Villiers and A. P Cronje had been minor ZAR officials before the war, but all of the five appointed had either not fought in the war or sided with the British.[131] The only one with any political base in the Transvaal was H. P. E. Janse van Rensburg, a prominent Heidelberg resident largely unknown outside the town. All five supported Chinese labour importation, although only van Rensburg organised a petition from his town to support the scheme and asserted that this reflected all Afrikaner feelings. It was the only pro-Chinese petition from a predominantly Afrikaner town in the Transvaal, although advocates insisted that Afrikaners in large numbers signed CMLIA petitions in circulation.[132] English-language newspapers supporting Chinese labour importation also made a point of referencing the Heidelberg resolution to 'prove' Afrikaner support.[133]

Such a situation highlighted the difficulty of 'proving' public opinion. Such claims could be made because of the relatively low profile Afrikaners maintained throughout 1903. When Chamberlain visited in January 1903 and the subject was first proposed, Smuts wrote a circular urging that 'we speak to our authorities with one voice as a people's party ... as it is most undesirable to arouse political ferment among our people, we request you to hold no meetings and to gather the views and feelings of your districts privately'.[134] This desire for privacy made it possible for Milner and others to emphasise early supporters and dismiss later pronouncements against the scheme as opportunistic attempts to divide the British colonial population, rather than a sincere reflection of 'public opinion'.

Despite this, the first large Afrikaner public meeting since the war was held in Heidelberg on 2 July 1903, organised by several former generals. A resolution written by Smuts, signed by Botha and put forward by Hendrick Alberts, the former Commandant of the Heidelberg Commando, stated clearly that

> the importation of Asiatics as a working class... will largely contribute to the closing of the Transvaal for white immigration... the employment of Asiatics... is not calculated to promote the permanent interests of the people of *South Africa*, and is therefore a measure which ought not to be adopted until such time as the white population of the Transvaal will be able to decide thereon under self-governing institutions.[135]

The resolution was passed by approximately 2000 Afrikaner men and duly sent to the colonial administration. Botha's eventual election as head of the *Het Volk* party in 1905, with Smuts in his cabinet, safely demonstrates that this was a fair reflection of at least a significant portion of Afrikaner opinion. However, while Milner forwarded it to the CO, he also included a covering letter dismissing the resolution as opportunistic: 'if they saw a chance of making political capital out of the opposition to imported labour', they would. Afrikaners, according to

Milner, recognised the divisions among the British and sought to exploit them for their own ends.[136]

Implicit to such an argument was a link between securing Chinese labour and securing the Transvaal as a British colony. The most widely read newspaper in South Africa insisted repeatedly that Chinese labour was the only way to secure the promised jobs for unemployed British, Australian, Canadian, and New Zealand soldiers still in South Africa and looking for work. Not getting Chinese labour would 'effectually prevent the Transvaal from ever becoming truly *British* in its white population'.[137] Similar comments were made in British parliamentary debates by those in favour of the scheme and by the Cape Progressives.[138] Those in favour of Chinese labour were also keen to play up the anti-capitalist nature of much of the opposition. The Cape *Telegraph* complained that 'the problem would have been solved long ago had it not been for the barrier of suspicion which seems to separate a large section of South African colonists from the mining magnates... There can be no more foolish or false suggestion than that the mines could obtain labour in the country if they tried'.[139]

While these interpretations of Afrikaner opinion were certainly clouded by prejudice, Afrikaners were themselves decidedly ambivalent about what approach to take. Smuts explained that Chinese labour importation 'presents much difficulties that we are still waiting events and opportunities'.[140] Similarly, Emily Hobhouse explained that 'the Dutch keep quiet and say "It is your business, fight it out."'[141] *Die Volkstem* even claimed that Botha advised fellow Afrikaners: 'the responsibility of the proposed step must rest entirely on the shoulders of those who want to import the Chinese'.[142] There was also a general feeling that there was no point 'appealing either to the Legislative Council or officials, which we do not recognize, or to Lawley or Milner'.[143] Indeed, the legislative council became known colloquially by Afrikaners as Milner's *debatsvereeniging*, or debating society, to denote their view that it was simply his mouthpiece.

After finally organising a petition to the British Parliament in February 1904, Smuts wrote dolefully: 'I do not know whether it [the petition] has done any good in England; it however served to clear the air here, as all sorts of reports were about to the effects that we were "neutral" in this fight.' However, even he acknowledged to Emily Hobhouse that there were reasons for this impression:

> That a large proportion of the Boers are apathetic is no doubt true; but these are the people who have lost all hope and heart;

who are prepared to see this Government do anything in the Transvaal...beneath this apathy burns in the Boer mind a fierce indignation against this sacrilege of Chinese importation – this spoliation of the heritage for which the generations of the people have sacrificed their all.[144]

The ORC with its large Afrikaner population was somewhat different, being the only 'white' colony not to issue any official resolution regarding the scheme. Like the Transvaal, the colony was run by an appointed legislature chosen by Milner and overseen by his administration. The population was largely rural and the devastation from the war was severe. Consequently, there was no widespread support or opposition to the scheme. Milner claimed 'the attitude of the majority of the people is precisely what I believe to be that of the Boers in the Transvaal, namely, that as long as they are not troubled with Asiatics living among them as farmers and traders they would heartily welcome any increase in the supply of labour, from the scarcity of which they themselves suffer acutely'.[145] He was probably somewhat correct in this belief, as several farmer associations discussed the matter and, while against it, hoped it would provide greater African labour supplies for their farms.[146] There were strong objections, however, most notably from Marthinus T. Steyn, the respected former president, who was committed to a united South Africa and wary of the mines becoming too politically powerful.[147] *The Friend*, an Afrikaner bilingual newspaper, warned that the Chinese would 'beget a bastard race which would spread' to the ORC,[148] while urging 'the two [white] races of this Colony' to work together to 'add their protest' so that 'it will be seen that South African opinion preponderates heavily against the Chinamen, and the Imperial Government will decree accordingly'.[149] The English-language *Bloemfontein Post* expressed deep reservations despite being part of the Argus group, owned by Corner House: 'The policy...is nothing but coquetting with "the yellow peril."'[150] A meeting of the Farmers' Union Congress, made up of local Farmers' Associations, 'objected to the importation of Asiatics into the Orange River Colony or Transvaal' and subsequently opposed 'granting Asiatics any trading rights'.[151] The former Transvaal Commissioner of Mines, William Wybergh, travelled around the ORC giving speeches and organising petitions and a few were sent, although not on the scale of the other southern African colonies.[152]

The general inaction in the ORC was partly because of their own labour shortages, partly because there was an informal promise of selling their produce to Chinese workers[153] and partly because they already had in place bans on any Asian immigration into the colony. Despite bordering Natal, a ban since the 1880s had effectively prevented any Asian from entering the colony, so they could better distance themselves from the fears expressed elsewhere in southern Africa.[154] Furthermore, like in the Transvaal, much of the Afrikaner population opposed to Milner's government were wary of their new imperial masters and preoccupied with the devastation from the war. Canvassing an imperial government they deeply resented and distrusted was simply not a priority.

Other colonies were far less reticent about expressing an opinion, especially in New Zealand. Throughout 1903, Richard Seddon, the longest serving New Zealand Prime Minister in history, organised popular rallies in New Zealand and Australia, which usually attracted several thousand people. He repeatedly emphasised the rights of white men to be self-governing, reminding a Wellington audience that 'each of the self-governing colonies had laws strictly prohibiting the Introduction of Chinese, and to flout these colonies by the Transvaal agreeing to their Introduction would be regrettable and cause heartburning and a revulsion of feeling.'[155] Seddon even wrote to the CO insisting that Australia and New Zealand should first conduct a joint enquiry to determine whether the labour was really necessary.[156] He also tried to organise a joint petition from all the self-governing colonies. Canada refused, the Cape already had, while Alfred Deakin, the Australian Prime Minister, was initially reluctant. This stemmed from his recognition that such interference in imperial matters created a 'new and grave departure' from the usual practice of self-government (he was no fan of imperial federation, unlike Seddon). Deakin was finally convinced to take this unprecedented step, however, after both Australia's and New Zealand's parliaments passed almost unanimous resolutions against the Transvaal scheme. The petitions and the parliamentary resolutions argued that, at the very least, such an important decision was not for an imperial government to make, only for a self-governing colony.[157] The parliamentary protests were passed on by their respective governors to Lyttelton and both prime ministers sent a pre-agreed telegram to Milner that they foresaw 'grave perils, racial, social, political and sanitary' for South Africa, 'however stringent' restrictions were and urged against the scheme.[158]

Canadians did not join the petition because they largely held that they did not have the right to interfere in a local matter.[159] The Canadian Prime Minister, Wilfred Laurier, had most of his support from within Quebec, a province with little interest in Chinese migration or in securing specifically 'British' colonies, nor were labour organisations powerful enough to make their campaigns against the importation politically significant. The government's refusals to interfere also reflected the sensitive nature of anti-Chinese feelings within Canada. Several times, the judiciary and central parliament had conflicted with British Columbia over control of immigration policy; the rest of Canada simply did not share its concerns regarding Asian immigration. Consequently, British Columbia independently sent a cable to Milner opposing the scheme, and some local protests occurred in Toronto and other large cities, but there was no widespread objection.[160] Even in Australia, there were people who thought it 'impolite to interfere with matters outside its jurisdiction'.[161]

Despite Southern Rhodesia's Legislature advocating the introduction of Chinese labour into their colony, there were petitions from various towns and prominent individuals such as the Bishop of Mashonaland, all against any Chinese importation. Indeed, the *Bulawayo Observer* even ran an editorial declaring that South African federation would be wrecked 'by the gratuitous introduction of a new Chinese complication'.[162] Robert Huttenback has suggested that the Rhodesians wanted the Chinese until 'the envisaged nightmare of hordes of Chinese laborers descending upon the Transvaal and then slowly insinuating themselves into Southern Rhodesia, considerably cooled the territory's ardor'.[163] It seems more likely that, like in the Transvaal, the Legislature's calls did not represent the views of the wider public. They never imported Chinese labour.

Natal's government chose a similar position to Canada's, although they formally cabled the CO to inform them of this. The *Times of Natal* complained: 'It amounts to this, that a purely *South African* question of the utmost moment – a matter which vitally affects the prosperity, even the very solvency of the country – is to be made a party cry in England.'[164] Furthermore, a large portion of Natal's income came from the goods shipped to the Transvaal mines; if the Chinese arrived at Durban and travelled through the colony to reach the mines, there could be a significant profit for the colony.[165] As Milner explained approvingly: 'In Natal, which itself depends for prosperity on such labour, there is a marked unwillingness to interfere in the controversy going on in the Transvaal.'[166] Newspaper coverage seems to confirm such a view, with most extremely imperialistic and pro-Milner,[167] such as the Natal *Mercury*, the most popular newspaper in the colony, edited by H. Ramsay Collins, a friend of Milner who acted as his trusted go-between when negotiating with Natal Indians.[168] Coverage focused on defending Natal's own right to decide upon immigration matters. Just as they had campaigned, and nearly rioted, to get their language test approved, so they equally felt it their right to decide whether to have indentured labour or not. The Legislative Assembly of Natal voted repeatedly against a resolution submitted by a private member prohibiting the importation of Chinese labour into southern Africa.[169] The Natal Premier forwarded their explanation: 'Natal holds the opinion that additional unskilled labor is urgently required for the development of the Transvaal, and is unable to join in the protest.'[170]

Of the three leading English-language newspapers, only the *Natal Witness* initially opposed the measure, as did a small Zulu-language newspaper, *Ilanga lase Natal*.[171] A final attempt to pass a resolution against importation gained only 2 out of 32 votes. Consequently, there were none of the petitions seen elsewhere. The Maritzburg Chamber of Commerce was almost unique in writing to Milner urging him to support the scheme.[172] By then, Natal had been guaranteed that Durban would be the arrival port for the Chinese and significant taxation would be paid to them so they wrote in support of the scheme.[173]

Transvaal responses

The Transvaal had decidedly mixed feelings about the attention the scheme received elsewhere. This is important to note because it emphasises the real differences which undermined efforts to form inter-colonial cooperation on the issue of Chinese migration. As previously mentioned, other self-governing colonies largely thought responsible government should be granted to the Transvaal first, so that the voting (white) population could decide for themselves. They were preoccupied with the rights of such a colony and the power Britain had over immigration policies. Much of the settler colony dialogue utilised the language of Britishness and 'white labourism', as well, debating whether Chinese labour would secure or destroy southern Africa's promise as a white colony, like Australia. Such debates fit in well with existing labour politicisation and ethnic patriotism. These did not necessarily fit in well with the Transvaal's own situation. Some questioned the assumption abroad that South Africa could ever be a white colony,[174] while others argued South African gold mining could never be run like Australian gold mining because unskilled labour was 'kaffir' labour, never white labour.[175] The Chamber of Mines was particularly keen to emphasise: 'This colony must decide for itself. It was no one else's business to interfere with them in their settlement of this matter.'[176]

Even among those against Chinese labour, some objected to other colonies interfering. There was a long-standing view that the real conflict in southern Africa was not between different races but was due to interference 'with the internal concerns of South Africa' from the British Government.[177] One 'British Colonist' who had lived in the Transvaal for 11 years demanded that Britain 'leave South Africans to work out their own salvation...self-government has become an essential part of the Briton's very nature...he chafes, becomes suspicious and discontented when he finds himself under even the mildest form of non-representative Government'. Neither Britain nor the other colonies could or should have a say; instead, he argued that the Transvaal should wait for self-government.[178] Others stressed that people outside of the Transvaal could only rely on hearsay, so depended on what they read, which could be manipulated.[179] To many, this situation was not an advertisement for formalising imperial federation.

Of course, others were grateful for this 'interference'. Between 50 and 60 Lydenburg citizens resolved that they 'view[ed] with alarm the deliberate attempt of the mining houses to saddle the mining industry of the Transvaal with Chinese labour' and referenced the way 'mines in America, Australia and Canada, of less average grade of ore' could successfully use unskilled white labour. If nothing else, they requested self-government be granted first.[180]

Referendum or self-government

While granting self-government increasingly became the argument of those against the scheme, the colonial and imperial governments repeatedly refused all such requests.[181] The Australian House of Representatives passed a motion that no decision be made until 'a referendum of the white population...or responsible government is granted'.[182] The *Aborigines' Friend* also thought this, as Milner's appointed legislature was under capitalist control and the scheme would be 'slavery'.[183] A Pretoria town resolution asked that the government 'postpone any decision and action in this matter until the mature views of the inhabitants of the Transvaal Colony have been expressed under a Responsible Government'.[184]

Among British Liberals in particular, previously negative or indifferent views of anti-Asian policies in the 'white' colonies was drowned out by a wave of anti-Chinese sentiment. Campbell-Bannerman, the head of the Liberals, even sought to pass a motion censuring the government for not automatically overruling the Transvaal legislature. Instead, he wanted Parliament to wait for the granting of self-government or a referendum to be held. The very nature of empire and the relationship between Britain and her colonies were at stake. He explained:

[T]he Imperial Parliament stands, as it were, in the place of trustee for the voteless and voiceless inhabitants of that colony. Had they with full constitutional representative authority put forward this measure and submitted it for the sanction of the Crown we might have thought they were in error, we might have disagreed with them and regretted their action, but we should have had to acknowledge that in their own affairs, and in their own House, it was not very easy or proper to interfere directly with them. But now, while they have no self-government, it is upon the Imperial Government and upon the Imperial Parliament, and, if we assent to the action of the Imperial Government, upon us, that the responsibility rests for a new departure, which may, if not checked, irretrievably influence for evil not only the future of the colony itself, but the whole district of the world in which it lies.[185]

However, the government majority held and his motion failed to pass.[186] Instead, the government repeatedly rejected pleas for a referendum or the granting of self-government: 'the Government are simply assenting to an Ordinance which has been exhaustively discussed by those responsible for the Government of the Transvaal, and approved of by the eminent statesman who represents this country there'.[187] Other colonies who objected were both interfering in something which was none of their business and were motivated by self-interest. In the official response to the petitions from Australian and New Zealand parliaments, W. E. Davidson, the Colonial Secretary for the Transvaal, sought to reassure them that white labour could never succeed in southern Africa, and that restrictions would be firmly enforced, while dismissing the protests as 'deriving ... from the conditions existing in Australia, which are very different from those prevailing in South Africa'.[188]

If the government really wanted to support the rights of self-governing colonies, they had to support the Transvaal Legislature's decision. After all, 'this Resolution was passed by nine votes to four of the unofficial, and by twenty to four of the official and unofficial members together ... a grave responsibility would be on our shoulders if, without cause or reason, and without any other remedy to suggest, we prevented the discussion of this measure in the Transvaal'.[189] Ignoring the Legislative Council 'would be a most dangerous principle'.[190] In the House of Lords, Lord Norton even threatened that interference might destroy the empire:

I can conceive nothing more disastrous to the interests of this country, both Imperially and from a colonial point of view, than that local concerns of our distant Colonies should be subject to the judgment of, and be made material of Party warfare, in the Imperial Parliament...there are still many leading men in this country who have a sort of itch for that meddling in colonial local government which lost us all our first great Colonial Empire, and which we are now, after a lapse of two centuries, invited to adopt in the case of our newest Colonies.[191]

Furthermore, if they waited for self-government, the colony would have to 'pass through a period of confusion and insolvency. Its finances are to be disorganised, its development is to be paralysed'.[192]

Both parties thus appealed to the rights of self-governing colonies to decide matters for themselves, and both argued that their rivals threatened the foundations of the empire. The Unionists argued that the Legislative Council and Milner's reports about public opinion in the Transvaal showed that the scheme should be approved. The government never cancelled local legislation without very good reason; to override the decision now might threaten the new colony, won at such cost. The Liberals, meanwhile, maintained that the Transvaal was against the scheme and, if real democracy existed in the colony, the measure would never have been passed. To them, it was largely a question of corruption: the government, and thus empire itself, was being corrupted by financial interests. Also, the mines were corruptly using their influence to skew perceptions of public opinion.

Corruption

This question of corruption suffuses contemporary accounts, and subsequent analysis has focused on accusations of official subservience, bribery and intimidation.[193] Indeed, such a view was firmly cemented within the anti-Chinese discourse during the initial White League's 1 April meeting, when speakers stressed that the Inter-Colonial Conference resolution in Bloemfontein was fixed by Milner, who had appointed docile followers rather than true representatives of public feeling. One anti-immigration speaker complained that the Conference decided the labour question 'in the people's name without the people having an opportunity of being heard'.[194] Fremantle, although editor of the *SAN* in the Cape, complained that 'the Press is systematically bought and blinded, and as far as possible a strict monopoly is exercised over the channels by which the exclusively English-speaking public, both of South Africa and of England, obtains its information and much of its opinions'.[195] In New Zealand too, newspapers said: 'the magnates, having throttled every expression of free opinion in the press'.[196] Even within the CO itself, A. B. Keith accused Milner of deliberately choosing evidence to prove his own view and leaving out other opinions.[197]

This suspicion of a government–mine conspiracy was merely an extension of suspicions voiced during the war, most famously by J. A. Hobson, but which became even more popular in the decades after the war had ended.[198] It is important to remember that ideas of 'economic imperialism' were not popular, in Britain or elsewhere, during the South African War. Peter Cain has argued that Hobson 'helped to establish, in the minds of anti-imperialists, the belief that the war was being fought at the behest of "alien financiers" who had hijacked the British state'.[199] But Hobson's publications remained relatively obscure at this time. The idea that the government's imperial policies were controlled by mining interests really only became popular as a result of the controversy over Chinese labour. Many had suspected the motives for the South African War; Chinese labour confirmed their worst fears. Rumours persisted that Chamberlain had organised a secret deal with the mines while visiting in January 1903: he would allow Chinese migration and the Transvaal would pay £30 million of the war debts, underwritten by the mines.[200] He publicly refuted this idea, going so far as to claim that such a deal would have been 'almost treasonably on my part',[201] but the rumours persisted. Chamberlain was, after all, an industrialist from Birmingham and had interests in South African gold mining. It was thus reasonable for the public to suspect that he would be more sensitive to mining interests than to other concerns, rightly or wrongly.[202]

These suspicions were exacerbated by mine company ownership of most of the major southern African newspapers. Contemporaries and subsequent historians have argued that the 'Chamber was able to ensure that an impression was created of widespread support for the proposal among the white (i.e. British) community in the Transvaal...through their buying power, patronage and ownership of newspapers',[203] regardless of 'real' public opinion. By 1902, all of the major southern African newspapers were dominated by three groups, of which the Argus Group was the biggest. The Argus Group was itself largely owned by Corner House, giving them control of the Johannesburg *Star*, the Cape *Argus*, Bloemfontein *Post*, *Bulawayo Chronicle*, and Salisbury *Herald*. The *Cape Times*, the *Transvaal Leader* and the majority of Kimberley's *Diamond Fields Advertiser* were owned by the *Cape Times* Company (with most shares owned by Corner House). The *Rand Daily Mail*, the Johannesburg *Sunday Times*, and the Pretoria *Transvaal Advertiser* were all owned by the South African Mails Limited.[204] Mining companies were undoubtedly the main investors in these groups and their newspapers often supported mining interests; the fact that these newspapers so assiduously supported British intervention in southern Africa in the run-up to war had increased the view that they were the mouthpieces of the mines.[205] Potter is right to point out, however, that these groups still had to compete with each other and with independent titles such as the *South African News* or the Pretoria *Volkstem*, which does limit any potential for actual conspiracy.[206] Furthermore, as Porter has observed, while editors frequently did favour the policies of their owners, this had more to do with initial hiring policies than 'influence' over newspaper content once an editor was hired.[207] When an editor, such as Monypenny at the *Star*, publicly opposed Chinese labour importation, he was replaced (although Monypenny always insisted he had chosen to step down because he knew his position would be against mine interests, rather than because any pressure was brought to bear on him).[208]

Undoubtedly, regardless of how much control mines had in the day-to-day editorial policies of newspapers, this did give them a strongly sympathetic media in many of the major southern African newspapers. This was especially true as many British and other colonial newspapers bought their stories from these newspapers, a scheme first devised to save money during the South African War. Combined with the propaganda efforts of the ISAA and the CMLIA, they had a powerful, and potentially misleading, influence on how 'public opinion' appeared. And they certainly were capable of outright misrepresentation to achieve their objectives. For instance, William Mather, a Cumbrian miner and founder of the TMA, initially one of the most vocal against the scheme,[209] spent the latter part of 1903 campaigning in Britain in support of it, pretending to still be a union leader (he lost his position due to financial irregularities).[210] He himself admitted that his change of heart had been 'financially "profitable"'.[211] Edward Perrow was another initial opponent within the TMA who then travelled around the Transvaal promoting Chinese labour in late 1903, and was subsequently given a comfortable job as an Inspector of Mines.[212] A. Jabez Strong travelled around Britain in early 1904 falsely claiming to be president of the Miner's Union and that Transvaal labour unions were in favour of Chinese importation.[213] It was widely believed that their support was bought. Certainly their former union, the TMA, held six meetings around the Rand in August 1903 against 'the unmanly action of the management of various mines in attempting to coerce employees into signing for importation'.[214] By February 1904, a Cape newspaper complained: 'In fear, the anti-Chinese party has crumbled away. There is no money for the opposition, even if any limited amount would fight so wealthy and absolute an organisation as that of the magnates and financiers.'[215]

The fact that both those for and against Chinese labour used petitions to 'prove' public support also made much of their evidence questionable. For instance, most of the Transvaal meetings which passed resolutions supporting Chinese importation on the Rand were held on mine property, where voting free from management interference was doubtful. These meetings were always overseen by mining managers in a position to dismiss their workers, or to otherwise bribe them.[216] The most clear evidence of some degree of corruption is the fact that the petitions and resolutions were worded almost identically and were passed on to the colonial administration by the CMLIA secretary, while most other petitions were sent directly to the colonial government by meeting organisers. The largest petition organised by the CMLIA had 45,078 adult male signatures; suspicions were raised because the entire adult white male population of the Transvaal was 90,000, while the Chamber also had to publicly deny workers were threatened with being sacked if they did not sign it.[217] In the Lichtenburg area the constabulary was accused of helping collect signatures.[218] Eventually senior mine managers had to publicly admit that some signatures were fake and later removed them from the list.[219] While it is impossible to know how many of the accusations against the mines are true, and while it is important to remember that not all the mines seem to have been involved in such practices, the false representations of the miners, Mathers, Perrow, and Strong, along with the dubious origin of the petitions, does indeed look like at least some mining figures were willing to do whatever it took to gain the appearance of public approval, which only added to suspicions of a mine-government conspiracy to import the Chinese.

Final approval

So with so much controversy, why did the Unionist government decide to support the importation of Chinese labour? Partly, it became a matter of party reputation. The Liberals repeatedly attacked the personal integrity of Milner and various government ministers involved in the decision, making them less inclined to back down. It also seems clear that they did not expect the British public to care as much as they did. Initially, the government decided to let the Transvaal Ordinance stand and merely explained that the economic situation in the Transvaal required the labour and the Legislature had voted for it. After the final

Ordinance was passed in the Transvaal Legislature, the Liberals had attempted to have a Royal Commission appointed to investigate the scheme before Parliament voted on the matter. The mines were actually willing to approve this, to show they had nothing to hide, but wished to avoid a delay if possible.[220] The Unionist government, eager to accommodate given the economic crisis, decided instead that proceedings in the Transvaal sufficed. The first (and eventually all) instalments of the War Contribution Loan were postponed at the request of Arthur Lawley, the Lieutenant-Governor in the Transvaal, due to 'the financial position here and at home'.[221] The British government accepted the argument that, if they wanted to be repaid, Chinese labour was essential.[222]

This reflects one of the dilemmas facing the Unionist government. Their intense focus on the economic necessity of Chinese labour both encouraged claims of capitalist influence while also tying the government to the plan, regardless of the political fallout. Once the government said the mining industry would collapse without Chinese labour, and that the entire southern African economy might do likewise, it became a matter of confidence. If they backtracked on Chinese importation, there probably would have been a stock market crash and any loans to the Transvaal government would surely be written off. As one critic complained: 'The mines had simply had too much invested... Chinese labour was an excuse to delay a market crash in France and Britain, and had nothing to do with the welfare of South Africa.'[223]

Furthermore, while Chamberlain had been deeply sceptical about public opinion in the Transvaal, he was no longer the Secretary of State for the Colonies when the decision was finally made. It is questionable whether he would ever have given approval. In the run-up to the 1906 election, he was publicly critical of the scheme.[224] As Phyllis Wheatcroft has shown, Chamberlain had an 'uneasy' relationship with Rand mine magnates and Milner after the South African War and 'One of his last utterances before the stroke which ended his public life was to accuse ministers of colluding over the Chinese question with "Magnates."'[225] He never gave up a belief that white labour would have worked while being convinced that Milner and the Unionists were too ready to believe arguments in favour of the scheme. His replacement since October 1903, Alfred Lyttelton, however, was a close friend and admirer of Milner, had previously been the head of the ISAA, and viewed Milner's strong support for Chinese labour without criticism. Indeed, in one telegram, he explained that the government 'could not sit still and do nothing till a decision could be taken under self-governing institutions'.[226] Chamberlain might have eventually conceded to the scheme, but he probably would have taken his time to be more certain of the necessity and of public opinion.

Ultimately, while Chamberlain had emphasised originally that public opinion was essential and that the Transvaal Legislature's vote would not suffice, the Unionists accepted the argument that, as the imperial government rarely intervened, they would allow the Legislature's Ordinance.[227] Consequently, the Secretary of State for the Colonies approved the scheme in January and the Unionist parliament concurred in February 1904. In this, the Unionists made a clear decision that only the Transvaal's opinion mattered and they chose to believe pro-Chinese lobbyists. As Lyttelton explained to Lord Northcote in Australia, Australians had every right to their opinion, but the government wanted to treat 'the Transvaal as a self-governing Colony so far as its internal affairs are concerned where Imperial interests are not involved, and the presence of active local support and the absence of active local opposition' made it impossible to turn down the proposal.[228] While many Liberals, and a few Tories, continued to oppose the scheme, the Unionist majority meant that Milner and the mines got their wish for Chinese labour. The politics of empire would never be the same.

Conclusion

Accusations of corruption did not stop the importation of the Chinese, largely because many of the proven cases of corruption were not exposed until 1904, when the scheme was already approved. Such incidents ensured that the matter would not rest simply because approval had finally been granted. It also fostered the general feeling that the British public did not know what it was talking about when it came to colonial matters, especially in southern Africa.[229] Such a situation estranged many southern Africans from the very idea of British involvement in their affairs and prevented them from seriously considering imperial federation favourably in the future. Many Britons and colonials had fought to help secure all of southern Africa for Britain. Their estrangement from British politics would have a lasting impact on the formation of responsible government in the Transvaal and on the union itself. As Merriman stated: 'I have yet to learn that the people of this country are to be governed by leading articles in the British press, or to have their affairs settled 5000 miles away by the people – irresponsible people – who write in the newspapers.'[230] By the end of the experiment, this attitude would predominate.

4
A Question of Honour: Slavery, Sovereignty, and the Legal Framework

The decision to import Chinese into South Africa was not merely an imperial matter, of course, but fed into national discourses in Britain and China. A belief that the mines and government colluded to secure Chinese labour has been widely reiterated since by historians. Richardson has claimed that '[n]o clearer evidence of this co-operation between State and industry can be found than in the history of the most important single piece of legislation in the whole experiment, the Labour Importation Ordinance'.[1] However, this is based on the assumption that the primary players were the Transvaal government and Chamber of Mines. Analysis of the establishment of an administrative framework for the importation of the Chinese reveals that this is far too simplistic. While there was a general desire within the Transvaal administration and the Chamber to cooperate to make the scheme a success, since improving the profitability of the mines was important to both, there were too many competing interests for them to work in harmony. The Chamber of Mines itself represented nine mining companies and over 140 mines in 1903 (see Tables 4.1–4.4). Furthermore, the Transvaal administration, the British Parliament, the CO, the FO, and different branches of the Chinese government were all involved in shaping and administering the terms of indenture and all under the watchful eyes of the public. Conflict was inevitable and collusion problematic.

In Britain, the 'Chinese slavery' controversy remains one of the most important examples of empire taking centre stage in popular politics, and received more newspaper coverage than the Congo 'slavery' scandal at the same time.[2] In the run-up to the 1906 British election, the use of Chinese labour on the gold mines of the Transvaal helped unite different labour and liberal factions, obsessed both with promoting the rights of 'white labour' and with preserving Britain's humanitarian reputation. It went to the heart of liberal notions of why the British Empire existed and challenged British identity as a humanitarian nation in the wake of the South African War. Consequently, the legal framework of the scheme became an important medium through which empire and British national identity could be debated.

The legal framework was also important because it was the first time the Chinese government was directly involved in drawing up the terms of Chinese indentured labour. The result was a contract which enshrined working and living rights which far surpassed the norm in South Africa, Britain, or China.[3] In China, the imperial government tried to use this as a means of reasserting their strength and appeasing the growing nationalist support within their diaspora. In practice, it revealed the deeply ineffective and fractious nature of Qing rule at this time. Terms were agreed in London, but implementation in China and South Africa was fraught with difficulties. Instead of demonstrating Qing strength, the scheme merely demonstrated regional power struggles, the growing support for Sun Yat-Sen in the diaspora and foreign economic imperialism in China. Furthermore, while negotiations focused on housing, wages, recruiting, remittances, health checks, corporal punishment, government oversight, and who paid for what, they failed to take into account simple things like language barriers or how to actually enforce all of the terms agreed upon.

Chinese 'Slavery' and British 'Honour'

The primary focus of negotiators in the Transvaal had been to mitigate comparisons with the large Indian 'coolie' scheme in Natal and to prevent the Chinese from competing with whites in skilled work, agriculture, or trading. In Britain, the view was different, despite the mine propaganda efforts. Since indentured labour schemes began to be used after the abolition of slavery, there were widespread allegations that this was simply a new form of slavery.[4] From the 1840s, the government frequently held commissions and published parliamentary papers on the investigations into the various Indian indenture schemes around the world, endeavouring to ensure good treatment. If widespread abuses were found, shipments could be stopped almost immediately. By the 1870s 'protectors' were normally appointed by colonial governments to oversee the fair treatment of indentured labourers. Nor was it just British officials who sought to improve conditions. Increasingly, Chinese and Japanese governments and the India Office monitored the treatment of their nationals abroad, even terminating shipments themselves when conditions were deemed poor.[5] Attempts to regulate the industry were also hampered by the isolation of most indentured labourers on their farms. Mistreatment would usually only be witnessed by other non-white workers, whose evidence in court was considered unreliable compared to white witnesses. Leaving the premises without permission, even to complain of maltreatment, was usually against the law.

The terms of indenture on this occasion seemed particularly open to accusations of 'slavery'. Already controversial in Britain, the experiment became the most debated indenture scheme ever in Britain. Focusing so exclusively on the 'slavery' debate could be used as a distraction from local issues, such as increasing divisions between Liberals and Labour, as Kevin Grant has argued.[6] Peter Cain has made a similar point regarding the general appeal of the theory of 'financial imperialism' at this time.[7] This was certainly an aspect of the attention on the Chinese scheme and explains its importance in the run-up to the 1906 British election. The issue tapped into 'the disillusionment of the country [begun] with the Boer War', that the government was being too closely controlled by capitalists and that British labour and British morality would suffer in consequence.[8] After all, Pro-Boers, who had believed either that the Boer republics should have been allowed to remain independent or that Britain's wartime conduct had been inhumane, were numerous in the Liberal ranks. [9] The issue crystallised genuine anxieties many Britons felt about the nature of imperialism, especially in southern Africa, and about the potential for national decline within Britain, threatened by competition from elsewhere. In a social Darwinist world, moral decline was to be feared as much as anything, and the turn of the century revealed many such anxieties in Britain. Appealing to a nationalism which took pride in Britain's 'civilising mission' was also a clever way to sidestep accusations that Liberal criticisms of government policy were unpatriotic.

The link between slavery and Britain's national honour was frequently employed by the Liberals to emphasise a link between government and capitalism. Herbert Samuel MP said 'the terms of their service approximate too closely to servile terms to be palatable to a nation which for a century has regarded its advocacy of universal freedom as among the chief of its glories'. Not only was this slavery, according to him, but he spoke of 'the discredit to the Empire that will ensue' as a result.[10] H. C. Thomson bemoaned 'how widely it has diverged from the humane principles of the English Law'. The scheme risked 'mak[ing] her bankrupt in honour' and might therefore 'shake the very foundations of our power in the Far East'.[11] Moral decline and imperial decline were easily linked. John Burns MP talked of the 'degrading spectacle' of 'the slavery of the yellow race, which is now proceeding with the approval of this gold-spangled Government'.[12] William Harcourt MP called it 'a degradation of the British name' and 'this demoralising mischief.'[13] Campbell-Bannerman MP, as leader of the Liberals, called it 'the biggest scheme of human dumping since the Middle Passage was abolished', referencing the Atlantic slave trade. It was a 'conscienceless and nauseous proceeding' which 'involves the negation of all the social, economic, and political principles which have given us and our Colonies our position in the world'.[14] John Fletcher Moulton MP remarked that

> [t]he whole fame of the name of the Empire is involved... this is a moment when England as a nation and as an Empire has to decide, for good or for evil, whether she will turn back to the bad ways of the past or raise one stage higher the standard of national morality and of the recognition, I will not say of the equality of man, but of the dignity of human nature.[15]

The government were taking 'advantage of the misery' of Chinese poverty rather than serving as a moral compass for 'less civilised' peoples.[16] The Bishop of Hereford, leader of opposition to the scheme in the House of Lords, thought the 'spirit of it is altogether un-Christian' and 'the essence of slavery', and even suggested Milner be recalled for supporting the Ordinance.[17]

These evocations of Britain's imagined glorious past were meant to emphasise the unsuitability and lack of ethics of the Unionist party to govern, just as it emphasised how they had allowed financial interests to place 'a black stain on the British flag'.[18] Cartoons, postcards, posters, and even poems on the manner appeared, such as:

> The slave whose wandering foot by Fate was led
> To British soil that very hour became
> A freeman. Dead, we thought, was England's fame,
> Dead every hope, if Liberty fell dead.
> But now the halo fades from England's head:
> We dally with dishonour. Huge our shame
> When the soul's prostitution we proclaim,
> Defiling lands where noblest blood was shed!
> ... now the yellow myriads we enslave
> And with their soulless toil dig Freedom's grave:
> Grasping at gold, damnation we attain.[19]

Frequently campaigners combined racial prejudice with concerns for Britain's humanitarian reputation, without any sense of contradiction. The accusations of slavery were intertwined with stereotypes of British love of freedom and morality, contrasted with Chinese immorality and African vulnerability. The MP, John Burns, summed up the views of many when he stated: 'it is to make a hundred Kimberleys – only worse – over South Africa, that this Asiatic labour is to be introduced, to the permanent harm of the Kaf[f]ir native, the injury and displacement of British workmen, and inevitably, by that process, to the loss of that Colony to the British Empire'.[20] Chinese labour threatened British jobs and it threatened the ability to make South Africa a 'white man's country'. It also went against the duties the British had towards less 'civilised' peoples, whether Africans or Chinese. It made their campaign in the Belgian Congo appear hypocritical and seemed to reflect the same imperial moral corruption evident in the South African War concentration camps.[21] This scheme foresaw 'the discredit of the empire'.[22] The references to 'honour' and 'slavery' made it easy for them 'to protect white skilled workers against "cheap" labour competition while wrapping this cause in a cloak of morality'.[23] It was the perfect issue to unite Liberals with the growing Labour and Socialist parties.

Such claims were not always consciously exaggerated accusations for political ends. If these had purely been opportunistic accusations, it would not have had the resonance across the country that it clearly had. For instance, on 26 March 1904, a Hyde Park procession of about 50–70,000 trade unionists and labour supporters marched against Chinese labour, one of the largest British protests ever made against an imperial policy. One widely photographed banner read: 'Slavery abolished in 1833. Restored in 1904 after the sacrifice of 52,000 men, women and children.'[24] Such public displays continued in the run-up to the 1906 election. One popular pamphlet, *Uncle Tom's Cabin* (the title itself a reference to the abolitionist movement in the United States) reflected the duality of the election campaign. At the same time as lambasting the 'slavery' of the scheme, it embraced Creswell's white labour solution and included a signed affidavit from a miner who claimed that he had been paid 15s. by mining magnates to create a disturbance and stop a proper vote on Chinese labour being taken during one Johannesburg meeting in December 1903. This and other examples were put forward to show 'the ruthless manner in which the mine-owners crush anyone who exhibits anything like independence'.[25] The democratic and economic power of white men was threatened by mining magnates, as were basic British humanitarian principles. Britain's racial supremacy, so linked to their moral superiority, was in danger of being lost. This was not entirely cynical, but was a fundamental tension of empire at the time, what Stoler and Cooper have called the 'tensions between the exclusionary practices and universalizing claims of bourgeois culture', a tension often evident in Chinese immigration debates.[26] These cries of 'slavery' and racial superiority and free labour were entwined in a powerful muddle of prejudice and moral righteousness, revealing the deep anxieties about national decline within British society in the early twentieth century. While there were contradictions in such appeals, but few Britons seemed to notice that the Liberals campaigned simultaneously for a protectionist white labour policy and the ousting of Chinese labour from South Africa while also pushing for 'free trade' (as opposed to the other great election issue, tariff reform). Unionist contradictions in policy fared less well. They were mocked for their Aliens Act (1905), which sought to prevent Eastern Europeans (mostly Jewish) from 'flooding' Britain, by excluding impecunious, diseased, or criminal migrants travelling third class (first and second class were exempt). It was the first legislation in the United Kingdom to control immigration at the point of entry[27] and was portrayed as yet another attack on free trade, which fit in well with Liberal claims that the Unionists were destroying a Britain which championed free labour and free trade, as well as morality.

Legislative adaptations

While Unionists made concerted efforts to counter such accusations with statistics, legal precedent, and economic 'facts', the outcry did lead to significant changes to the terms of indenture. Unionist claims simply 'did not have the same sensational appeal that the Liberal cry of "slavery" had'.[28] While the Liberals failed to halt the scheme, their references to 'British honour' and the widespread public support for such a view did force the government to scrutinise the Transvaal Ordinance as few other pieces of colonial legislation had ever been scrutinised before, and this had a marked effect on the scheme, something which had not been planned for by legislators in the Transvaal.

Indeed, their own parliamentary negotiations had been long-winded so the bill presented to the Imperial Parliament was already a strange mixture of clauses. Most significantly, a labour member, Raitt insisted his approval was contingent on the ordinance including Farrar's list of skilled and semi-skilled jobs which the Chinese would be banned from doing. Against legal advice, and without asking mining colleagues, Farrar agreed.[29] Farrar was keen enough to get the experiment underway to agree to these highly restrictive terms, which he hoped in future, once their scheme proved successful, to overturn. By specifying appropriate jobs so precisely, the mines would be far more severely limited in how they utilised Chinese labour than they even had been with African labour. Within a single day's negotiations, the basis for South Africa's infamous industrial colour bar, which restricted certain types of labour to certain racial groups, was put in place. And unlike the British government's clear objections to race-based legislation concerning Asians, here they made little effort to stop the legislation.[30]

In Britain, many Liberal proposals were practical attempts to mitigate what they saw as some of the most unfair or 'morally dangerous' aspects of these regulations, although it is perhaps significant that there was no widespread Liberal attempt to remove the ban on the Chinese undertaking 'white' skilled work, despite its racism, as this would have been unpopular with many British voters. One significant change had to do with the transfer of employees from one mine to another. One of the most frequent complaints from Africans was that a mine could choose to move them to another mine without their permission. Due to parliamentary pressure, the regulations were changed so that labourers had to consent to any transfer or any other change in the terms of their contract once they were in southern Africa and families would always be transferred with the workers. As Lord Monkswell bragged, this condition 'was put in originally, struck out at the instance of the mine-owners, and has been put back again at the instance of the British people'.[31]

Some liberals actually chose to support the scheme in exchange for specific alterations. For example, the Archbishop of Canterbury and Archdeacon Furse of Johannesburg made it clear that their approval was conditional on women being imported to prevent 'the moral evils that might result from such immigration'.[32] Their close scrutiny of the Transvaal Legislation was unusual, with the Archbishop of Canterbury speaking at length on the matter in the House of Lords. Both made clear that wives were needed or prostitution would become rampant and their mixed-race offspring would become a permanent feature of the Transvaal.[33] This combination of concern for the conditions and well-being of the Chinese and a belief that the Chinese were particularly lascivious and immoral were typical of criticisms of the scheme. Most indentured schemes which allowed families to migrate together required the entire families to work. Because of the political pressure, the government forced the mines to agree that wives and children under ten could be imported at the expense of the importer but would not work. The mines would be required to provide married accommodation (but not food).[34] This was a substantial bow to those who feared what the results of Chinese sexual appetites on the southern African population might be. Fundamentally, this was a provision that would never have been made if not for the concerns about British moral authority and honour and surpassed the treatment of married European miners, who suffered from a shortage of married accommodation. In return, the scheme advocates had secured the support of two of the most high-profile Church of England figures.

In other cases the mines voluntarily changed the terms, in order to allay public suspicions. Given that company share prices now depended on the scheme, Alfred Beit acknowledged from London: 'It will... be necessary, both in the interests of the mines and of the Chinamen, to see that they are properly housed, fed and looked after' because 'the moral aspect of the case will certainly receive considerable attention from this side of the water.'[35] Indeed, as parliamentary debates dragged on, many in the mining community were now willing to increase expenditure to ensure public and government support. Despite the additional costs involved, the mines willingly agreed that, if a Chinese labourer was injured at work 'not through his own fault or carelessness', the employer had to pay 50 shillings and if a man died the employer had to pay 10 sovereigns to an appointed 'representative' and for the funeral, 'according to Chinese custom'. The employer would even have to pay for the body to be returned to China, if the worker desired it.[36] Most skilled white miners received no such promises from their employers.

Similarly, when some of the mining magnates wished to adopt a closed-compound system similar to that used on the Kimberly diamond mines to control the Chinese, Corner House directors in London convinced their Johannesburg colleagues to opt instead for the normal Transvaal system of an open compound, both to appease the Transvaal merchant community and to counteract slavery accusations in Britain. Also, while existing Rand miners typically bought food outside the mine premises, and cooking facilities were rarely provided, the CO insisted that the regulations laid out specific rations. This would ensure that the Chinese were fed properly and that this was evident to the public.[37] Consequently, the terms were revised to include the provision of two hot meals a day with unlimited tea and rice, and meat in each portion. The food was to be prepared by 'free' Chinese imported alongside the 'coolies', although suppliers would be local when possible.[38] The Chinese would be entitled to Chinese-style food and at the expense of the mines. Each mine had to build housing which, for the first time in southern Africa, required a set amount of cubic living and working space for each man. No such accommodation existed, so the mines had to construct suitable buildings quickly. The 'Chinese compounds' where the Chinese would live would also have hot and cold water, a luxury not found in white accommodation, married or otherwise. In addition to this, each mine had to open a new hospital under a white doctor with Chinese assistants and provide Chinese doctors and Chinese medicine; labourers could choose between Western and Chinese medicine, an additional expense meant to aid recruiting efforts and appease public concerns.[39] Lyttelton insisted, despite strong mine protests and 'the great expense which would be involved', on a surgeon to oversee health checks in China before being shipped to Durban and independent Transvaal government supervision over recruitment and all other aspects of the scheme.[40] Again, none of these provisions were available to European or African miners at the time.

While most historians have focused on 'the grim conditions endured by these Chinese labourers in the Transvaal gold-mines' as an example of 'the callous, wasteful, inefficient and brutally exploitative ways in which the gold-mining companies treated their unskilled labourers at this time',[41] the regulations theoretically represented a significant advancement in living conditions. In winter, the Chinese received extra blankets, the living quarters were often described as 'light and airy...They are in every case superior to what the labourers are accustomed to in China'.[42] While there would be significant problems with the scheme, it also 'led to an improvement in housing conditions, for minimum standards were laid down for the Chinese, and when they left these quarters were used for Natives'.[43]

Some aspects of the scheme remained largely the same as outlined in the Transvaal Ordinance. The Transvaal would be required throughout to provide independent government supervision (at mine expense), which eventually resulted in the Foreign Labour Department (FLD), established in March 1904. William Evans, formerly Protector of Chinese in the Straits Settlements, was the first Superintendent of the FLD. Evans had been of particular help to Skinner during his travels and had subsequently advised the Chamber about recruiting conditions and the Chinese 'character'.[44] His inspectors were mostly British Army soldiers stationed in China or Hong Kong, many of whom had helped suppress the Boxer Rebellion.[45] In practice, he and his inspectors were meant to supervise the mines and the Chinese, ensure they followed all regulations and take them to court or fine them if they broke any regulations.[46] This was in addition to the department's responsibility for overseeing all aspects of administering the scheme in China, the Transvaal, and en route between the two.

All of this oversight was to be at the expense of the mines. Traditionally, indenture schemes were set up and funded either directly by the local colonial government or by private business groups. In most colonies, one-third of expenses were paid for by local taxation and the rest by private organisations.[47] However, at the insistence of Richard Solomon, the Transvaal's Lieutenant-Governor and Attorney-General, it was agreed that they paid all importation, repatriation, and supervisory expenses, and any other incidental expenses connected with the scheme.[48] Solomon saw the desirability of the Chinese economically, but as a Transvaal official he was primarily concerned with ensuring that the scheme did not rely upon local taxation.[49] Milner's reconstruction was costly and the government needed to cut expenses wherever possible. In this case, the mines had to accommodate the government.

Despite all of these compromises, there remained serious problems with the negotiations. As one Liberal peer complained, 'the modifications made in the Ordinance have been forced on the Government by public opinion that I feel there is lacking that guarantee of good faith and sincerity that would have been desirable'.[50] The governmental and mine willingness to alter the terms of the Ordinance did not guarantee that those terms would be enforced with any rigour. While the government had insisted on expensive and bureaucratic levels of oversight, they had still appointed Evans, knowing him to be on friendly terms with the Chamber and possibly biased in their favour. Indeed, while there was clearly a desire to make the scheme look better to a British and Transvaal public, not all modifications were practical or sufficiently specific. For instance, while expenses were to be covered by the CMLIA, there would be intense debates subsequently about whether aspects such as policing were covered by the license fees or whether the mines should pay an additional sum to the colony.[51] Furthermore, the approval in Britain was for a modified ordinance, which was never legally approved in the Transvaal, leading to a legalistic minefield.

Negotiations with China

Adding to the general muddle of the legislative framework was the position of China in negotiations. From the beginning of negotiations, it was clear that any scheme would require the cooperation of the Transvaal, British, and Chinese governments, and would be based on the terms of the 1860 Peking Treaty between Britain and China.[52] However, in the past, the Chinese government had never been formally consulted about indenture, so the matter was given little consideration. Furthermore, no one seriously considered the problems inherent in conducting negotiations between so many people and organisations over such great distances. A single point of wording could easily take several days of expensive cable exchanges before an agreement could be found. Indeed, it was not until 11 February, two weeks after Milner had cabled the Transvaal Ordinance, that the Chinese finally offered their first official response.

The recent anti-Chinese activities of the 'settler' colonies made the Transvaal request for labour different from previous schemes. While settler colonies at this time had increasing political independence, they still had no power over foreign policy. This had not mattered so much previously because Europeans simply recruited through the Chinese ports they controlled unofficially. However, this time the head of the Wai-Wu-Pu[53] told port officials repeatedly that while 'the engagement of coolies' was permitted by Treaty, they were not to allow any Transvaal recruitment until the terms China sought were agreed to.[54] Furthermore, China refused to negotiate directly with the Transvaal, because it was not a sovereign state. It had consistently denied the rights of colonies to restrict their subjects and now was their chance to publicly snub colonial governments. Richardson has even suggested that the Chinese government was trying to prevent any future 'coolie' schemes by insisting upon requirements they never thought any potential employer would accept.[55] While there is no direct evidence to support this, it is plausible that the Chinese government at the least wanted to make an example out of the case, and did not care whether the scheme ever went ahead. The Chinese government was particularly keen to be seen to be protecting its subjects' interests at this time because of the increasing pushes towards 'modernisation' within China after the Boxer Rebellion and the political activism of middle-class Chinese communities abroad who wanted protection from regulations and prejudices which targeted them. Previously, 'the Chinese Government has too many troubles at home to protect the interests of its wandering sons abroad'.[56] Recently, however, the Japanese government had been far more proactive and the Chinese government wished to emulate them to some extent.

Consequently, when the local Transvaal administration approved the importation of the Chinese in December 1903, the British government had to conduct all negotiations with a hostile Chinese Consul in London. While outright refusal would have been difficult, the Chinese Minister, Chang Ta-Jên, could publicly make a point of emphasising why the Chinese were desired as labourers and their mistreatment abroad, while demonstrating the might of the Chinese government to 'protect' Chinese subjects. While he thanked the FO for consulting them on the matter, he hit back: 'A similar course on the part of some of the self-governing Colonies who have passed laws offensive to the Chinese Government would have obviated the necessity for much of the acrimonious correspondence from this side.'[57] In other words, the Chinese government clearly saw this as an opportunity to restore national dignity. The humiliating treatment during the Boxer Rebellion was a factor but perhaps even more important was the treatment of their migrants in the settler colonies. This is clear in the pointed demand from Minister Chang that all negotiations had to be between Britain and China, with no 'cognizance of any such person as the "importer" mentioned in the Ordinance' or of the colony itself.[58] The diaspora had previously been ignored by the Chinese government but the leading financial supporters of Sun Yat-Sen, the future leader of the 1911 revolution, lived abroad in the United States, Australia, South Africa, and Britain.[59] The Chinese in South Africa had tried for some time to improve their situation, one of their reasons for supporting people like Sun. They had made frequent complaints about their treatment in the colony.[60] One of their most insistent demands was the appointment of a Consul in the Transvaal to oversee the welfare of the Chinese in South Africa.[61] The London Consul now demanded 'the appointment of Consuls or Consular Agents to watch over the interests and welfare of the immigrants'. This Consul would have the right to visit the Chinese and their places of accommodation 'at all reasonable times' and to report on his findings to the Chinese government, in order to ensure their fair treatment.[62] By demanding the introduction of a Chinese Consul, they were theoretically going far beyond the conditions of any previous schemes and making a point of appeasing the local Chinese population in southern Africa.

While the mines did not want a Chinese Consul involved,[63] Lyttelton thought it impossible to disagree with the appointment if Chinese approval was to be given, pointing out that 'the arrangement could be reconsidered' at a later date.[64] It was agreed finally and the CMLIA would even pay for his house and other expenses, but with the actual power of the Chinese Consul kept vague.[65] He did not have the right to visit the mines without permission to check on their conditions, had no real power of enforcement, but could report on the matter to China should he wish to do so. As the head of WNLA, F. Perry, unflatteringly explained: 'The last thing we want is a petty mandarin, with an army of subordinates, registering the individual labourers and certainly blackmailing them in the process.'[66] The *Star* later described the man from Chilli eventually appointed as 'a Mandarin of the second-class... an important personage in his own country... He is accompanied by his daughter and some mine officials...His arrival in Johannesburg will doubtless be a potent influence for good amongst his fellow countrymen on the mines'.[67] This different view was sparked by the man's Christianity. In reality, his actual contact with indentured workers appears to have been almost non-existent. In other words, this was a free way for the Chinese government to give the permanent population the Consulate they had long wanted and had little to do with concern for indentured labourers.

In addition to the bickering over the Consul, the Chinese wanted corporal punishment to be illegal and treated as 'common assault', the importer to be the employer, not merely a go-between and for repatriation to be to the port of embarkation, not merely a return to China. They also required a minimum wage be specified. Such terms were unusual and would add greatly to the expense of the scheme.[68] Most of the terms were agreed to at a London meeting[69] but the matters of a minimum wage and corporal punishment proved to be two of the most contentious issues of the entire scheme. Both became a point of national honour for both the British and the Chinese. To protect the Chinese and to assure the public that they were not being imported to lower African or white wages, the Chinese Consul and Lyttelton insisted a minimum wage was included despite strong Transvaal objections.[70] While most indentured schemes specified a set wage, this had been left out of the Transvaal Ordinance because the Chamber had long since decided, upon the advice of their experts, to pay the 'coolies' by piecework rather than fixed daily wages whenever possible. This was not the practice with Africans, but racial stereotyping was all-important in this decision. The head of WNLA, when investigating labour potential in China, found 'the Chinaman of the coolie class is a most slack and inefficient worker if paid by the day or month, but that a great change is effected by paying him according to work done'. It might also encourage him to 'work considerably longer than the standard hours, thus increasing their relative value'.[71] Not all work could be carried out by piecework, however, so writing into the Ordinance the wage terms was problematic. Lyttelton finally approved a compromise which specified the average wage to be paid after six months as 50 shillings per 30 days, or 1s.6d. per day; also no Chinese would be forced to do piecework against their will.[72] However, it was left unclear how an 'average' was to be derived. These negotiations dragged on from February to May, significantly delaying the start of the scheme and epitomising the problems of the negotiations. The Transvaal's own Attorney-General called the clause about wages 'the most badly drafted and unintelligible clause in any contract I have ever had to consider'.[73]

The issue over corporal punishment was equally convoluted. The Chinese were not successful in having flogging banned outright, as

> any labourers imported would be amenable to the law of the land by which everybody, including whites, is liable to corporal punishment for certain offences. But no corporal punishment would be allowed except for such offences, and in these cases could only be inflicted after trial and sentence by a Magistrate or Judge.

The instrument used had to be government-approved, the lashes could not exceed 24 for adults and a medical officer had to confirm that the prisoner was fit for flogging before the punishment could be carried out. With such reassurances, the Chinese did not press the point.[74]

In practice, this is not what happened. In 1905, it became clear that the High Commissioner, Deputy High Commissioner, and Superintendant of the FLD, meant to protect Chinese interests, all sanctioned flogging by mine supervisors similar to that used in English schools, in order to maintain discipline among the Chinese, and without any real oversight on their parts. Milner later claimed that the approval of such punishment was neither secret nor meant to break any rules but 'it seemed to me so harmless, that I really gave very little thought to the matter'.[75] However, he never informed the CO of this policy decision. While, within the Transvaal gold mines, enforcing a ban on flogging would have been difficult, the CO, especially under Chamberlain, had been trying to eliminate the practice, as Milner knew.[76] From 1902, colonies had to send the CO reports on all flogging cases. And Chamberlain even gave a speech on the subject when visiting South Africa in January 1903. In the presence of Milner, he declared that 'if it is suggested that a Kaffir must be flogged for a breach of contract etc., then I protest against it as being contrary to the English character, unworthy of the English, and inhuman'.[77]

While it is clear that flogging workers was standard practice in many British colonies, including southern Africa, it was unusual that Milner explicitly sanctioned extra-judicial corporal punishment at the request of the mines. The question thus remains why Milner and Evans as FLD superintendent agreed to the mines' request. The files are vague about this issue as there are no records to indicate exactly when the policy was adopted. The mines probably successfully argued that to have to go to a magistrate every time a Chinese labourer did something wrong would be impractical. It was a common practice of South African whites to utilise beatings to control their African labourers, and was particularly common on the mines.[78] There had never been serious public outcry or government interference regarding the existing system, so they failed to consider public scrutiny or that they were breaching the British Convention with China. And while Chamberlain in parliament called it 'an error of judgement' on Milner's part, he did defend his actions, as he thought it was all at Evans' instigation.[79] If there is a single case which highlights an overly close relationship between the mines and the senior colonial administration, it is this decision.

It also reflects the general ineffectiveness of the Chinese government to enforce any of the terms of the indenture once the scheme had begun. The significance of the negotiations between London and China in the Transvaal was limited. The Transvaal administration deliberately undermined the terms relating to corporal punishment, and may have done so in other cases. Furthermore, the Chinese show of strength was clearly short-lived. The Chinese Consul appointed in Johannesburg attempted to investigate abuses only once in 1906, but the mines objected to his involvement and there is no evidence that he attempted to interfere in any other cases, whether through ignorance or indifference.[80]

Recruitment

The problems of applying the negotiated terms were especially evident in recruiting. A detailed recruiting infrastructure in theory was negotiated between the different parties. William Perry, head of WNLA, the African recruiting agency for the Chamber, was sent to China to begin organising recruitment in December 1903.[81] However, some within the Chamber thought WNLA incompetence had led to the present labour shortage; others simply thought someone with more Chinese experience was needed.[82] Consequently, in the spring of 1904, the CMLIA, originally set up to organise public petitions supporting the scheme, was placed in charge of recruiting in China. The shareholders of this non-profit organisation were the mines themselves, all members of the Chamber, who contributed £500,000 in total towards the operations and with the power to increase their funds up to £1 million if necessary, a considerable financial commitment, especially for the smaller mines. The Board members were the same as on the Executive Committee for the Chamber. Of the 140 mines within the Chamber, 40 signed up to actually receive recruits (see Table 4.2).

To manage the CMLIA, Farrar successfully convinced the other mining magnates to appoint Major Walter L. Bagot.[83] Bagot had fought in the Nile Expedition in 1898 and in the South African War, gaining medals and promotion, before securing a position within Farrar's company,

ERPM, and was clearly Farrar's man. The CMLIA had their headquarters in Johannesburg, as well as staff in China at each of the depots used.[84] J. A. Brazier was appointed as general manager of the CMLIA in China, and Butterfield and Swire and The Chinese Salt Organisation were hired to recruit and transport labourers.[85] Some were, like in the FLD, ex-army or civil service although many were from the small merchant communities in China and some of the lesser positions went to 'free' Chinese or 'half-castes'. The CMLIA also hired staff required by the Ordinance: a European Overseer who spoke Chinese (dialects were unspecified and consequently varied, rarely in relation to the actual Chinese recruited), medical officers and cooks.[86]

The FLD was represented by Transvaal Emigration Agents (TEAs), who were there to oversee all recruitment and ensure the terms and conditions were enforced in each depot, numbering three in 1905 (Hong Kong, Chefoo, and Chinwangtao).[87] They were also to ensure that all recruits were voluntary.[88] There were also numerous other staff. For example, the Surgeon Superintendent, responsible for medical checks, had to be 'a British subject' and qualified medical practitioner, as agreed by the British and Chinese governments, although appointed by local TEAs.[89]

As with most British indenture schemes involving the Chinese, it was imagined that most recruiting would be done in southern China. However, problems quickly arose in Amoy, Hong Kong, and Canton so most recruits actually came from Shantung or Chihli and travelled out of the ports in Chinwangtao and Chefoo. Many explanations were given for the failure to recruit successfully in southern China but largely it reflected the fractured nature of Chinese governance at this time and the competing interests of foreign powers. Several FO men claimed that either the Germans or Japanese were publishing false stories in order to undermine recruiting efforts throughout southern China. One placard in Amoy told locals that the Transvaal was hot, the 'inhabitants as black as pitch [and] go stark naked'; malaria killed most immigrants and 'venomous reptiles and wild beasts' killed the rest. Furthermore, the placard alleged that workers were kept in cages and punished with 'an iron whip'.[90] In another case, a Chinese man named Lo Cheuk in Hong Kong was sent before a British judge for trying to 'dissuade certain intending emigrants for South Africa to run away' from Butterfield & Swire's office. He allegedly told recruits that the death rate on the mines was very high 'while they were also told they would be engaged to serve in the war, and that they were sure to die'.[91] Local suspicions were made worse by the requirement to photograph and fingerprint each recruit, a measure normally reserved in China for criminals.[92] Butterfield & Swire, in charge of recruiting in southern China, paid for 'three village "headmen" from up-country' to visit the Transvaal to better inform locals about the real labour conditions, but the damage to the reputation of the scheme in southern China was already done.[93]

Worse still, the southern Chinese provincial authorities were uncooperative, even in Hong Kong itself. They had not been consulted when the terms were drawn up, and many were already involved in recruiting for schemes elsewhere (even British).[94] The Viceroy of Kwangsi (now Guangxi near Vietnam) demanded that only he could appoint recruiters in his province, and so five members of Butterfield & Swire staff were imprisoned for a short period for illegal recruiting. Even a Maritime Customs clerk was charged with recruiting men already in the depot awaiting deportation for a rival Mexican scheme. One persistent rumour claimed local 'mandarins... went round their district gaols promising all the inmates a free pardon if they would go to work in South Africa. As it appears a large number of the prisoners expected death as the result of their misdeeds, it is easily understood that there was no lack of volunteers'.[95] One ship report stated that there were 'a small proportion of men who, from brand marks, the absence of queues, and from other indications, appeared to be of a criminal type'.[96] In Wuchow, local officials apparently sent groups of prisoners still in chains as potential recruits, undermining recruiting efforts. Perhaps unsurprisingly, the FO banned any recruiting in the district after this became known.[97] Chinese officials in Canton were also uncooperative, demanding that they personally choose recruiters and that the recruiters work for the local Chinese government, 'not Hong Kong merchants', but all at the expense of the Transvaal. Different Chinese officials also fought internally to try to restrict deportation to specific ports. For a time, the Wai-wu-pu and Chinese Consul in London even claimed deportation from anywhere but Canton would not be allowed. Overall, there was clearly little official support for the scheme in southern China: 'The negotiations of the Convention in London...was regarded...by the high provincial authorities as a slight, for they hold that they should have been consulted in the matter.'[98] One magnate even asserted: 'the Chinese Government [have] a handle which they are only too ready to seize to bring about delay, more particularly if influential persons concerned in the matter are bringing pressure to bear upon the Government in order, in some way or other, to feather their own nests'.[99]

Overworked FO officials were also frequently unhelpful, especially as there was some resentment that recruiters were not made part of the Consular Service; in the past, 'emigration fees' were given to the FO for overseeing recruiting. FO officials were officially discouraged from interfering and some, like Robert Hart in Peking, outright admitted they did not 'want to be drawn into the matter. I have too much work of my own on hand'.[100] The lack of a clear authority figure to conduct matters further hampered matters, with CMLIA and TEA staff unable to organise diplomatic efforts locally because the Chinese government did not recognise their authority.

The first shipment from Hong Kong highlighted the recruiting problems. Three Chinese died during the voyage, several more upon arrival, and 40 immediately had to be repatriated to China as a result of beriberi.[101] Even three months later, a Hong Kong shipment included 61 with beriberi,[102] forcing the recruiters to stop their efforts entirely in the region.[103] By February 1905, 37 had died in South Africa from beriberi, almost all from the first two shipments out of Hong Kong.[104] There were also many repatriated: 13 returned for reasons such as 'lunatic', 'idiocy', 'fits', or 'imbecile'. One repatriated man was described as 'blind-insane', and a further six were simply blind. Several were also clearly too old for physical work.[105] All of these deficiencies should have been noticed in China. Theoretically, the Chinese should have been medically checked three times: once before formally signing a contract of indenture, once before the ship actually left China and again upon arrival in Durban, along with receiving smallpox vaccinations before arrival. This was an especially important point because of the supposed link between Asian migrants and disease. Preliminary medical examinations in China were the responsibility of the CMLIA, with TEA oversight, while the FLD oversaw the checks in Durban. Despite the seriousness of the matter, Perry was left complaining about 'the general chaos and confusion of duties and responsibility which the regulations in their present form seem designed to create'[106] and it is clear medical checks and vaccinations remained inadequate throughout the period.

As a result of the strong feelings on the matter, the CMLIA almost exclusively recruited in northern China after the first two shipments and quietly dropped the south altogether in 1905.[107] Subsequently, Chinwangtao became the 'centre of emigration', with Chefoo (Yantai in Shandong) close behind.[108] Of the 63,695 Chinese sent to South Africa, 97.3 per cent came from Northern Chinese ports, and of these, about 93 per cent were from the Chihli or Shandong provinces.[109] The incursion of Russia into Manchuria, the Boxer Rebellion, and continued drought greatly destabilised the region. As Richardson has noted, 'it is no accident that the most successful recruiting grounds for South Africa were exactly those areas which had experienced the greatest intensity of Boxer activity'.[110] Furthermore, new European mining interests in Chinwangtao in 1901 rapidly transformed it into one of the leading Chinese ports for British businessmen. The Chinese Engineering and Mining Company shared many shareholders and directors, including Carl Meyer, Percy Tarbutt, and the American engineer, Herbert Hoover, with South African gold mining companies, and their control of such a large company in Tientsin was credited with 'virtually saving Transvaal emigration'.[111] Perhaps most importantly, sending off the poor to South Africa proved a useful salve for local Chinese governments and Chinese men themselves.

The other major problem with the implementation of the legal framework in China related to remittances. The mines had decided to pay all Chinese an advance as a recruiting incentive and also offered subsequent remittances to families in China.[112] Each Chinese recruit was issued with a book, to be given to their families, who could then cash payment slips from the book once a month at a bank. However, many recruits were not accompanied by their families to the recruiting centres, so they depended on recruiters or friends to pass on the booklets. The CMLIA also lacked adequate access to banking or other organisations to reliably transfer money to relatives in rural areas. In December 1904, the Vice-Consul at Tientsin noted: 'The allotment system is a fruitful source of trouble and fraud...many of the books fall into wrong hands, though recruiters argue in favour of the system that it brings in a good type of men, chiefly of the agricultural class, who go abroad to help their families.' Chief abuses included handing over 'their allotment books to their creditors to satisfy their debts'. He even heard some sold 'them for fifty or sixty dollars paid down, being probably ignorant of their real value'. The reason for these problems given by the Consul was typical of the attitude adopted during subsequent difficulties: it was the Chinese character.

> Generally speaking, the allotment system affords peculiar facilities for the petty and elaborate system [of] blackmail so dear to the Chinese mind, and I have no doubt that it greatly commends itself to village headmen, official underlings, and the numerous sharp-witted rogues who prey upon the luckless coolie.[113]

Nine months into the experiment, a Transvaal Emigration Agent (TEA) offered a more sympathetic observation:

> Unless a coolie can entrust his money to a friend...from his own village, probably at great expense and inconvenience, the only course open to him is to hand it to the recruiter who brought him...He

must trust to the good faith of the latter for the remittance to reach his home. He has no safeguard whatsoever.[114]

Another TEA dismally observed: 'the allotment books are bought and sold and gambled and stolen and extorted to such an extent that it is impossible to say to what extent the right payees receive the money... It would however need a large detective staff to get the truth in this matter'.[115] No one was prepared to contemplate such a vast expense, even if they should have received the permission of relevant Chinese officials. However, given that some 10,000 families in northern China supposedly received assistance from relatives through remittance, and given its attractiveness to potential recruits, abandoning the scheme was unthinkable.[116] The Chinese government especially approved of the system and dismissed any suggestion of eliminating it. The issue was never resolved and surely would have caused greater problems if recruitment had not been cancelled in 1907.

Part of the problem was that few of those involved in the creation of the scheme had any practical experience in China. The CMLIA and FLD were almost entirely reliant on recruiters. These were mostly Chinese touts working inland who then took recruits to port cities in groups of 25–100. In 1905, the FLD investigated 'whether Chinese labourers receive any money to defray the expense they incur in reaching the port of embarkation from their homes' in order to encourage their travel, but 'in no instance could I discover that any money had been paid for this object; they had all of them been collected at a point of assembly (such as Tientsin) and from there had been paid their steamer or railway fares to the port of embarkation'.[117] Basically, the Chinese might easily have been travelling weeks before arriving at one of the FLD/CMLIA port depots and neither the FLD nor CMLIA had any control beyond their depots. They consequently had little control over recruiting methods or remittances, despite the regulations.

Language problems constantly added to the problems and was perhaps the most serious omission of negotiations. Because everyone was expecting to get the bulk of labourers from southern China, all but one of the FLD staff, including Evans, were from Hong Kong or Singapore. Many could read, write, and speak official Chinese but were particularly ill-prepared to communicate effectively with northern Chinese.[118] Even in China itself, this was a problem. Until October 1904, the TEAs at Chinwangtao 'did not themselves speak the Northern dialect and had to depend on the somewhat unreliable medium of a Chinese Interpreter'.[119] This was a serious problem indeed. It was the job of the depot staff from the CMLIA and the FLD to correct any misinformation given to them by Chinese recruits.[120] Chinese interpreters were consistently considered unreliable, while 'white' interpreters were generally too expensive and difficult to recruit, except for the most senior positions. While clearly there was a racist element to this diagnosis, it seems likely that the complexities of the legal framework were only poorly understood by recruiters and recruited alike. Each recruiting depot initially drew up its own contracts for the indentured workers and the specifics varied considerably. One drawn up by Butterfield & Swire for Hong Kong promised incorrect rates of pay and working conditions, as well as assuring Chinese audiences that the Chinese workers 'shall be treated as any other British labourer', something which the Ordinance terms clearly contradicted.[121] The one first used in Shantung and Chihli only described the work as 'on the gold mines' and failed to specify the work limitations laid down in the Ordinance.[122] Inaccuracies in the contract were usually noted when copies of the contracts arrived at the FLD offices in Johannesburg and they tried to standardise Chinese contracts to little effect. One Chinese interpreter, Gim Ah Chun, told a visiting British MP that the men under him at the Comet Mine had been recruited 'under the most cruel false pretences'.[123] Those who were unhappy either caused trouble and were repatriated, or came up with ingenious ways to be sent home. In one particularly dramatic case, a Chinese man purposefully mangled his fingers to render himself unfit for future work.[124] Other times, FLD inspectors, when quizzing newly arrived Chinese labourers, found they were 'cheerful and eager to get up to the mines and begin work'.[125] This made it exceptionally difficult to determine how willing the Chinese were to do the work they were signing up for.[126] Regardless of the truth, the publicity given to such cases ensured that the experiment 'confirmed' the ideas of kidnap and coercion which were so often associated with indentured labour, and thus undermined the purpose of the negotiated terms of indenture.

Conclusion

The final Ordinance inevitably bore the traces of the dozens of drafts and the numerous people involved in its editing, as well as the disparate aims of the interested groups. Furthermore, the negotiations reveal the importance placed on the scheme by elements of the British and Chinese populations. It became a question of power and of national 'honour'. Edward Rose in a popular pamphlet expressed a common shame that the only reason it was illegal to flog 'the Chinese Uncle Tom' was because the Chinese Minister had demanded it. 'If you wish to plumb the depths of turpitude into which the Randlords have plunged

Table 4.1 Numbers of Chinese labourers imported annually

	1904	1905	1906	1907
Yearly Number of Chinese imported	23,517	27,016	11,039	2123
Gross Total Chinese imported by 31 December each year	23,517	50,533	61,572	63,695
Number employed at 31 December each year	20,918	47,217	52,889	53,828
Loss	2599	3316	8683	9867
Loss (%)	11.05	6.56	14.1	15.49
Yearly percentage of unskilled labour imported	15.49	17.03	7.38	1.41
Percentage of total unskilled labour	13.78	32	35.36	35.86

Source: *Annual Reports* of the Transvaal Chamber of Mines, 1904–07.

Table 4.2 CMLIA Membership: Numbers of Chinese imported per mining group

Group	1904 # Imported	1905 # Imported	1906 # Imported	1907 # Imported	Total # Imported
Consolidated Gold Fields	3249	2967	2267	440	8923
ERPM	4321	6669	3740	306	15,036
General Mining	1916	1799	–	–	3715
Goerz & Co.	1929	1592	–	–	3521
J. B. Robinson	1988	–	–	–	1988
J. C. I.	3056	1847	496	–	8399
Neumann & Co.	1810	1440	208	34	3492
Wernher, Beit & Co.	5248	10,702	4328	1343	21,621

Source: *Annual Reports* of the Transvaal Chamber of Mines, 1904–07; FLD122–124; FLD343–53; HE252/136/591.

old England, ponder over that ironic fact. Chinese teaching England the elementary duty of humanity! – A heathen race rebuking a Christian Nation!'[127]

However, implementing the scheme would prove problematic, largely because of the unwieldy bureaucracy which the regulations outlined. While this made the regulations both innovative and ineffective, the nature of the compromises would have a lasting impact upon the

Table 4.3 Distribution of all labour by 'race' on the mines

Year	Producing Mines Whites	Producing Mines Africans	Producing Mines Chinese	Non-Producing Mines Whites	Non-Producing Mines Africans	Non-Producing Mines Chinese	All Mines Whites	All Mines Africans	All Mines Chinese
1903–1904	94,885	53,215		2499	8164	865	11,984	61,379	865
1904–1905	11,831	67,901	16,836	2704	8080	3060	14,535	75,981	19,986
1905–1906	13,874	70,224	38,429	3206	9189	3927	17,080	79,413	42,356
1906–1907	14,576	78,481	46,380	2102	7106	2496	16,677	85,587	48,876
1907–1908	14,714	103,158	30,590	2064	7993	1700	16,678	111,151	32,290
1908–1909	17,350	134,268	10,171	1453	7034	71	18,803	141,302	10,242
1909–1910	19,717	143,822	1862	2378	10,650		22,095	154,472	1826

Source: *Annual Reports* of the Transvaal Chamber of Mines, 1904–1910.

Table 4.4 List of Mines utilising Chinese labour

ERPM
Cason, 1905–10, producing
Angelo, 1905–10, p
Driefontein Consolidated, 1906–08(?), p
New Comet, 1904–08, p

NKG (Primarily Farrar and Anglo-French controlled)
Benoni, 1904–08, non-producing
New Kleinfontein, 1904–08, n-p
Kleinfontein Deep, 1904–08, n-p
Apex, 1905, n-p
Chimes West, 1905, n-p
Rand Klipfontein, 1905–06, n-p

Wernher, Beit & Co. and Rand Mines Ltd
Durban Roodepoort Deep, 1904–09, p
Geldenhuis Deep, 1905–10, p
Glen Deep, 1904–10, p
Jumpers Deep, 1904–10, p
Nourse Deep, 1904–06(?), p
Rose Deep, 1905–10, p
South Nourse, 1904–07, n-p

Consolidated Gold Fields Ltd
Simmer & Jack West, 1905–09, n-p
Jupiter, 1905–09, n-p
Simmer & Jack East, 1905–10, p
Simmer & Jack Proprietary, 1904–09, p
South Geldenhuis Deep, 1907–08, n-p
Simmer Deep, 1907–18, n-p

General Mining & Finance Corporation
Rand Collieries, 1905–09(?), n-p
Van Ryn Estates, 1904–09, p
Roodepoort United Main Reef, 1906–09, p
Violet Consolidated West Rand, 1905–06, p
West Rand Consolidated, 1905–08, p
Aurora West United, 1904–06, p

S. Neumann & Co.
Wits Deep, 1904–09, p

A. Goerz & Co.
Geduld Proprietary, 1904–08, n-p
Central Geduld, 1904–08, n-p
North Geduld, 1904–08, n-p
Lancaster, 1905–08, p
Princess Estate, 1904–08, p
Van Dyk Proprietary, 1905–08, n-p
Tudor, 1904–06, n-p

Table 4.4 (Continued)

Johannesburg Consolidated Investment Corporation
Consolidated Langlaagte, 1904–08, p
Roodepoort, 1904–07, p
Witswatersrand, 1904–08, p
New Rietfontein Estate, 1905–06, p
Van Ryn Deep, 04, n-p

Corner House (Eckstein & Co. and Wernher, Beit & Co.)
French Rand, 1904–09, p
New Herriot, 1905–09, p
New Modderfontein, 1904–10, p
Village Deep, 1905–10, p
Modderfontein Extension, 1905–06, n-p
Henry Nourse, 1905–06, p
Turf Mines, 1905, n-p

J. B. Robinson Group
North Randfontein, 1904–08, p
Robinson Randfontein, 1904,

p – productive
n-p – not productive

scheme and the relationship between the interested bodies. The mines had been forced to compromise, to accept restrictions they did not want and at great expense. The governments had also compromised and occasionally rushed through issues without seeking proper legal advice about every change made. Nor were the regulations, however involved, necessarily suited to the situations which would arise once the Chinese were actually imported. The costs involved had also spiralled steadily, putting more pressure on the scheme to succeed. The situation ensured that negotiations did not eliminate many of the objections already expressed in Britain, China, and elsewhere. The controversy's high profile globally was a crucial reason that the negotiations became a matter of honour, but the negotiations between the Transvaal administration, the British Parliament, the CO, the FO and different branches of the Chinese government, as well as the different mines, also reveals how important competing local concerns and networks remained.

5
Sex, Violence, and the Chinese: The 1905–06 Moral Panic

The violent beginning

Public disquiet about Chinese labour initially died down in the Transvaal after the Chinese began arriving in July 1904. A large section of Transvaalers depended on the mines for their livelihoods, and there was a somewhat grudging wait to see if the economy recovered as promised. There was even some tolerance for initial unrest, dismissed as teething problems. Most of the mine-owned newspapers promoted such a view. As late as March 1905, the *Star* tried to claim that there had been 'a few insignificant disturbances' but otherwise the Chinese were 'an undoubted success'.[1]

Attitudes, and eventually newspaper reports, gradually changed when accusations surfaced that many of the Chinese recruited were sick, insane, freed criminals, Boxers, or other 'bad characters'. The accusation of Boxers was particularly worrying, and difficult to counter, given that most recruits were from Shantung, the centre of Boxer activity. Any association with Boxers suggested the Chinese might be prone to violence towards Westerners. It fed not just opposition to the scheme but those interested in the sensationalist, exotic character of the Chinese. A popular British critic shared a common belief that the Chinese labourers included 'Scoundrels, professional gamblers, thieves with hands showing that they have done no manual work, "Boxers," and assassins'.[2] The savagery of the Boxer attacks, the authoritarian nature of Chinese governance, their wiliness, fatalism, and superstitions: these were increasingly invoked in public discourse and added to the already great interest in the scheme.

In addition to the initial recruiting problems, between July 1904 and July 1905, 28 separate riots involving the Chinese were reported to the FLD and many more violent incidents almost certainly went unreported.[3] Cases only had to be reported when the police were called to restore peace. Many cases of desertion, theft, assault, and murder were also reported widely in the press, in addition to the riots. From the summer of 1905 through 1906, this interest became an actual 'moral panic' in the Transvaal.[4] If one were merely to base an understanding of the scheme on newspaper reports, administrative files, and personal correspondence, the scheme would appear unusually violent. Almost all writing about the Chinese included the word 'outrage', although the majority of the hysteria, unlike most panics, depended more on rumour than on press reporting.

While there was violence, the rates did not match the public hysteria. Statistically, the Chinese were less violent than Africans; white violence towards Africans or Chinese was rarely recorded, and never in a systematic way. As well as distorting rates of violence, existing research fails to explain why the Chinese were viewed so frequently in terms of violence, nor has it been connected to similar Black Peril fears at the time. Like other panics in colonial Africa, sex and violence were frequently linked together and overshadowed other aspects of the scheme, even their economic purpose. The moral panic which arose in 1905 was not simply an obsession with rape, however, as was usually the case with Black Perils. Instead, it combined an obsession with Chinese sexual habits with a tendency to view the situation in terms of violence, predicated on long-standing racial, gender, and class assumptions about Africans, Chinese, and whites. Much of the discourse essentialised Chinese men as 'bad characters', violent, highly sexualised, amoral, and fatalistic in their approach to death, while white men were violent to protect their families, employees, or the racial hierarchy itself. Perhaps the most important reason for the dominance of sex and violence in discourses about the scheme was simply fear. White anxiety about racial miscegenation and insecurity about political or economic developments made Chinese labour a natural focus for colonial anxiety, especially as the Chinese were already a conspicuous and controversial presence in the colony. Once actual problems did arise, and in a region undergoing concurrent Black Peril panics in the Transvaal, Southern Rhodesia, and Natal (not to mention the wider Yellow Peril), it is not surprising that the Chinese seemed to threaten white values and interests. As earlier chapters made clear, the Transvaal was very unstable during this time, and despite some hopes, the Chinese did not bring security but came to symbolise insecurity to many. Nor was it difficult to portray the Chinese in a stereotypical manner, given the preceding coverage of the scheme.

This fear was exacerbated because the violence reflected the general fact (and feeling) that neither the mines nor the government were really in control, or worse still, that the government did not rule in the public interest at all. The discourse surrounding violence and sex also linked to popular aspects of Orientalist literature; the excitement, the titillation of such obsessions was a factor in the intensity of the focus on the Chinese. Overall, this discourse demonstrates the intersection of two types of moral panic: imagery from Black Perils from colonial Africa and yellow perils from the wider 'white world' were applied to the specific local context of the Chinese in the Transvaal.

Moral panics and performative violence

In any study of moral panics, there is debate over the degree to which moral panics are 'grassroots' movements or 'elite-engineered' to exert influence and control. 'The grassroots model sees moral panics as a direct and spontaneous expression of a widespread concern and anxiety about a perceived evil threat.'[5] The Salem witch trials in 1690s Massachusetts are usually given as an example of this kind of panic.[6] Others have argued that, in most cases, moral panics are about 'enabling the ruling stratum to maintain its privileged position'.[7] Dane Kennedy has adopted this approach in his work on white culture in early twentieth century Southern Rhodesia and Kenya, arguing that colonials deliberately manufactured and used Black Peril scares to gain greater rights from the government and to unite the new immigrants with older white settlers. He argues this is why they bore no relation to actual events.[8] Jock McCulloch, on the other hand, discounts the claim that 'Black peril was invented for political ends…There was no conspiracy'. Instead, fears over the threat of black men raping white women were very real, even if the threat was irrational, and even if those fears were still 'used to political advantage by white men and women'.[9] David Anderson presents a hybrid argument: some acts of violence were ignored because at the time or place they occurred, they were not needed. There were real rapes or murders, and the white population was genuinely vulnerable numerically but a panic only ensued when politically expedient.[10] All of this underlines the continued ambiguity within the terminology of the difference between a real and an imagined threat: 'the difference between actual and perceived threats is problematic and to distinguish between the two requires careful historical research of a kind which is only now beginning'.[11] In the Transvaal, while there were real and often horrific acts of violence, similar acts elsewhere or at different times did not receive the coverage that the actions of the Chinese received, nor cause the panic. Similarly, the sex lives of the Chinese were discussed, often with an openness and a frequency which bore no relation to the frequency of prosecutions for rape or homosexuality. This is what makes it a disproportionate panic.

Certainly, the yellow peril panic in the Transvaal cannot be dismissed as a 'conscious strategy'. The media, especially, did not consciously set out to generate a moral panic. Indeed, most local newspapers sought to downplay fears except during August and September in 1905, when violent assaults were at their peak. While political figures did seek to make capital out of unrest involving the Chinese, the actual panic did not begin until August 1905, after a series of high profile riots and murders in the space of a couple of weeks. It is important to note that there was a 'real' basis to fears, however exaggerated they became.

Analysis of the panic makes clear that both the actual acts of violence and the public panic became symbolic of wider issues of power. Stephen Ellis, in writing about Sierra Leone, has argued that people 'perform acts of violence'. Everywhere, he argues, violent acts contain symbols to those around them which signify specific social and cultural things within that environment, such as symbolising masculine virility and power.[12] The violence of Chinese assaults came to symbolise wider settler anxieties and fears. In Kenya, David Anderson has argued that violent punishment of Africans could symbolise settler power over Africans.[13] The violence directed at the Chinese can be understood in a similar way. White miners pushed for greater rights from the government, and increasingly campaigned against Chinese labourers altogether; assaulting the Chinese was a way to show the Chinese and the government the power they held, or wished to hold, regardless of the law, while the failure of the mines and government to control the Chinese was writ large as symbolising their indifference to the plight of 'real' South Africans. It could also symbolise increasing concerns about 'poor whites', most likely to be linked to sexual relations with the Chinese and with over-abusing the Chinese. Even the very act of sex could become symbolic in such an environment. Gareth Cornwell has argued:

> In the patriarchal construction of the sexual act, whether forced or not, the male is dominant and the female is subordinate: the political scandal of the Black Peril is the subjection of a woman of the dominant race to the power of a man of the subordinate race; the penetration of a white woman by a black man is an act of insurrection.[14]

It was for white men to set moral values, and Chinese sexual practices were thought to threaten those values. Even the Chinese might have been motivated by the symbolism of violence. Violence was one of the few avenues available to them to assert power over others.[15] Of course, given the archival information available, only speculation about Chinese motivation is possible. Whatever their reasoning, though, sex and violence dominated the scheme in a way which was never planned.

Economics and the press

So why did a moral panic occur in the Transvaal? Historians such as Charles van Onselen, Norman Etherington, and Jeremy Martens have linked similar Black Peril crises in southern Africa to periods of economic instability; such Perils were displaced projections of that instability.[16] Whites became obsessively aware 'their safety depended upon the maintenance of strict standards of behaviour among the African population'.[17] Part of the reason for the Chinese panic was also economic, although it fails to explain why economic anxieties should focus on the Chinese specifically. From mid-1905, the economic situation in all the southern African colonies certainly worsened. For the first year after the arrival of the Chinese, mining company shares had risen and the mines had made a point of starting new projects which increased the white labour force. On 31 December 1904, 16 mining companies employed 20,918 Chinese; on 31 December 1905, 35 companies employed 47,217 Chinese.[18] The output of gold had increased by 28.65 per cent. 5,555, while the number of stamps increased from 5,555 to 6,930 over the same period.[19] However, by the summer of 1905 the number of whites employed on the mines stopped growing; from October, it actually began to decrease, leading to unemployment. From early 1906, banks began to recall loans and in February mining shares were at a post-war low. Bankruptcies increased.[20] An unsubstantiated rumour in July 1906 that 'the Chinese cost the colony £100,000 per annum in increased policing, prisons, etc.'[21] did not help. Many thought the situation demonstrated that the legislation meant to control the Chinese was proving ineffective, especially when false rumours spread that the 'Chinese are now opening stores to trade with the mines in almost every district along the Rand.'[22] The labour union, the WTLC, complained to the colonial administration:

> At present the distress and poverty on the Witwatersrand is appalling, in Johannesburg alone there are hundreds out of employment, some of whom are absolutely starving, and yet withal, the Coolies ... are

slowly but surely being instructed to take the white man's place.[23] This belief was strengthened when mining figures showed that, while there was an increase in the numbers of whites working overall in 1905, the ratio of white to 'coloured' labour had proportionally decreased the number of whites employed. In 1903, there had been ten whites for every 59 Africans; in 1906, there were only 7 whites for every 59 Chinese and Africans.[24]

This served to generally heighten already strong anti-Asian sentiments, with Asian businessmen increasingly the subject of protest.[25] In Johannesburg, there was actually a decrease in Chinese business licenses between 1904 and 1905: there were 223 Chinese and 218 Indians with licenses in 1905, from 235 Chinese and 302 Indians in 1904.[26] The public press, however, linked the arrival of free and indentured Chinese, portraying them as working together to oust the white man. Asians generally were increasingly singled out in the press. For instance, normally Transvaal newspapers only put the type or location of a crime in their headlines. However, from the first arrival of the Chinese, they always indicated if crimes were committed by Chinese men, but often failed to clarify whether those involved were 'free' or 'indentured'. By the end of 1904, newspapers as far away as Port Elizabeth and Bloemfontein tended to specify whether a crime had been committed by a Chinese or an Indian, sometimes simply saying a 'coolie' had committed the crime, further blurring the division between different groups in the public imagination. It also made the issue of Chinese labour seem far more relevant, or at least of sensational interest, to widely diverse groups of people spread throughout southern Africa. Instead of the theoretically racial threat caused by 'John Chinaman' to labourers or white racial supremacy, it demonstrated a 'real' physical threat to residents.

Lack of respect

Economic resentment and skewed newspaper reporting alone cannot explain the resulting panic, although they helped to frame an environment in which such an event could occur. Anderson points out that most discourse in Kenya focused on 'insubordinate attitudes' by Africans rather than fears of 'criminality'; 'Africans simply were not subservient enough.'[27] Similarly, a frequent complaint was that that the Chinese were not respectful enough. On the mines, this meant that Chinese labourers were perceived to fight back, to attack whites, far more often than Africans. As one observer wrote: 'I believe the Chinese, who call us foreign devils, have more contempt for the white man than the black man has.'[28] This attitude largely spread because of the willingness for Chinese workers to attack white miners who mistreated them. After one fireman beat a Cheng Fang Chih so much, he ended up in hospital; other Chinese on the compound 'set upon the fireman, who was only rescued through the intervention of some other white men. All the coolies then struck work and marched off to the Compound'.[29] This was far from an isolated case, but white miners perceived it to be much rarer for Africans to retaliate in such a way. When the Chinese did this, it was usually described as Chinese 'insolence'. This did not always need to be violent resistance either. Yang Cheng Kuei was twice imprisoned for verbally 'insulting' white officials.[30]

It is also clear that such Chinese 'insolence' or violent attacks were not always limited to the mines, perpetuating concern throughout the Transvaal. One Greek storekeeper was cutting his opium before selling it on to the Chinese; in retaliation, 'his hands were lashed to the top of his bed, and his two feet to the bottom rail. Then a pad soaked in chloroform was held roughly over his mouth, and he knew no more... When he came to he... had been frightfully mutilated'.[31] Such reports combined the excitement of gothic horror with a genuine alarm at the potential power vengeful Chinese men could have. As such, commentary could focus as much on complaining about Chinese attitudes as on the actual violence which occurred. In another case, a disturbance broke out at an Indian shop, reportedly because some Chinese felt they had been cheated. The Indian hawker was killed and a 'kaffir' was then beaten to death, apparently for getting in the mob's way during their rampage. There were no witnesses forthcoming and no evidence to charge any individuals – the feeling that 'all Chinese men looked the same' prevented convictions and added to public discontent. In frustration and lacking any other clear motives, the Acting Superintendent of the FLD, wrote:

> This appears to be one of their characteristics and I very much doubt whether anything can be done to prevent it. If anything goes wrong, instead of appealing for assistance, they boil up in rage, take the law into their own hands and settle the matter for themselves. When the disturbance is over they appear to be very sorry for what they have done, express their regret and are willing to pay for any damage...[32]

Frequently, Transvaal Indians and whites alike complained that the Chinese would often use rickshaws and other facilities marked specifically for whites, something the local Asian population were not allowed to do.[33] The Chinese simply did not 'know their place' within the Transvaal's racial and class hierarchy. A Cape correspondent for the *Indian Opinion*, summarised local press coverage:

> yellow labour is a curse to the land in many ways... Instead of bringing prosperity and happiness to the people, it is enriching the rich and bloated mine magnates to the detriment of the interests of the mass... to-day stagnation and ruin are staring everyone in the face and the Chinamen are rebellious and contumacious in their misery.[34]

Even the simplest matters could lead to accusations of Chinese disrespect. The Compound Manager at the Simmer East Gold Mine, Germiston, decided that his Chinese police should wear old British army uniforms, widely available after the war. Yet a flood of criticism poured in from army officers still in southern Africa and various members of the public. In one case, a Royal Engineer actually forced a Chinese man to the Charge Office and took away the uniform. The problem, he claimed, was that 'it is detrimental to a soldier of H. M. Regular Army to see Chinese dressed the same as himself'.[35] The government subsequently required that the practice be banned.[36] More acceptably, through analysis of photographs and postcards, it is clear that many Chinese police wore the uniform of the Wei-hei-wei Regiment from China.[37] The regiment had developed a good reputation during the Boxer Rebellion, when, under British command, they had helped put it down. Disbanded shortly before recruiting began, it appears that many became police for the scheme in the Transvaal and continued to wear their Sikh-like uniforms.[38] It is worth commenting that no mention was ever made of these uniforms in South Africa, indicating that, while the significance of the British Army uniforms was recognised, the Wei-hei-wei uniforms were not understood by locals.[39]

The concern over the lack of respect supposedly shown by the Chinese was exacerbated by particular concerns that such an attitude might infect Africans. The Zulu rebellion in neighbouring Natal in 1906, although never directly blamed on Chinese labour, fed into the fears that the Chinese would somehow corrupt the 'native' population. One newspaper explained: 'The fear is that the two classes will spur each other on to a diabolical rivalry which will end in such gross atrocities that the world will be shocked by the revelation. Police and regulations seem less effective than tissue paper.'[40]

Attacks on the public

Undoubtedly the most important factor in the development of the moral panic was the Chinese attacks on white farmers in rural Transvaal. The change from a general anxiety to a full-blown panic happened in the summer of 1905 as a direct result of the murder of an Afrikaner farmer by housebreaking Chinese men in August 1905, popularly known as the 'Moabs velden tragedy'.[41] Piet Joubert was an Afrikaner farmer at Moabs velden in the Bronkhorst Spruit district. On 16 August 1905, he was stabbed and his wife, ten year old son and infant son were all wounded in the attack. Two daughters aged thirteen and fifteen witnessed the attacks from their room next to the living room. At around 11:30 pm, they were woken by noises and saw Joubert fighting with a Chinese man by the window. There were at least three other Chinese men, one with a pickaxe. One girl slipped out the window to grab nearby relatives. The other went to her parents' bedroom to warn their mother, but found the mother and baby hurt and laying on the ground outside. The older son was later found with a head wound. Neighbours from up to five miles away gathered and formed an armed search party, arresting the next day four bloodied Chinese found nearby with stolen Joubert property. The Reuters report spread the story around the world; it was the most reported case of violence involving the Chinese. An mid-American newspaper headline was typical of the coverage: 'TRANSVAAL FACES A YELLOW PERIL. Chinese Imported for Mines Have Become a Menace to the South African Peoples.'[42] Newspapers around the 'white world' began to cover regularly any crimes thought to involve the Chinese, revelling in the gory details, such as when twenty Chinese attacked an 'Indian's hut' and 'hacked the inmates about with knives'.[43] That said, there was a sort of weary, 'what can you expect' aspect to the coverage in places like Australia, which coloured their coverage.[44] Back in the Transvaal, local schools were even closed in some areas due to concerns for public safety.[45]

Settlers did not manufacture the panic which resulted from such attacks, but settlers could use these panics to criticise the colonial government. Southern Rhodesians used panics to criticise the government and 'the failure of the state to protect white families'.[46] In Kenya, panics were used to push the imperial government to grant more rights to the settler communities and to further disenfranchise Africans.[47] Similarly, the 'yellow peril' panic increasingly centred around criticisms of the government for failing to protect them, or for allowing the experiment to begin with. Several Afrikaner deputations to the High

Commissioner and Lieutenant-Governor during this period demanded compensation and better protection from Chinese deserters, with most requesting they 'be repatriated as soon as possible'. Even the mine-sympathetic Johannesburg *Star* warned that, if the government did not fix the Chinese problem, an Afrikaner rebellion might result.[48]

Most criticism focused on the issues of rearmament and compensation. A deputation of 21 Afrikaners, including Botha and Smuts, met the Lieutenant-Governor in September 1905, after the Joubert murder, to appeal for rearmament. One of the delegates, Wolmarans, reported that 'the people outside are living in a state of great unrest' and wanted 'the Government to send all the Chinese out of the country as soon as possible'. Until that could happen, they wanted to set up community vigilante groups to protect neighbourhoods and asked 'for compensation in those cases where damage has been done to people living outside'. Hiekman, from Pretoria, also complained that the reconstruction being undertaken on most farms mean that 'these places offer no resistance where four or five Chinamen wish to break in'. Liebenberg from Ward Crocodile River spoke of a murder perpetrated in his neighbourhood and, although no culprits had been caught, the Chinese were widely blamed. The government's protection was 'really quite insufficient.... Some hundreds of Chinamen are still wandering about the country... there is a feeling of intense unrest among the people so that on many farms people are taking out their windows and filling the places up, and at night they go to the other farms to sleep together for protection'. Because most did not have their own vehicles, they 'have to walk long distances morning and evening for this purpose'. He suggested they were willing to form a Commando to hunt for the Chinese but would need arms first.[49] Dieperink, the delegate from Krugersdorp, reiterated this view:

> I live about 800 yards from the Tudor Gold Mine where there are 3100 Chinese. There were three, now there are four policemen, a number altogether insufficient to protect the district. The state of affairs is such that for the last few weeks some of my children who sleep in outhouses, have had to sleep inside the house for protection, because we are quite unarmed and I think it is only right I should give my own experience, as I have been living for 11 years on the borders of the Transvaal among Kaffirs, there was never such a state of unrest as we have now on the Rand. The state of affairs is such that many of the work people have left the outside districts and gone to live in the town for the protection of their families... we have taken on us

the status of British subjects and as such we claim to be fully entitled to protection. The state of affairs is such that unless a change comes soon, something serious may happen.[50]

That warning was accompanied by reference to various recent cases in the Krugersdorp area, such as the murder of a 'Kaffir', and complained that the press seemed to deliberately not report such cases. The failure of the press to always cover murders did not calm the situation. Instead, it simply fanned the conspiracy theories over newspaper control.[51] One mining engineer told the *Illustrated London News* that 'the policy of the mine managers has been to hush up all trouble. The full story of the rioting has never been told'.[52]

While most moral panics emphasise the important role of newspapers in spreading panic, it is important to remember that in this case the fact that there were few official lines of information allowed rumour and fear to spread unimpeded by fact. Because people largely relied on gossip to spread stories, the fear, the suspicions of the government and mines, and the general gruesomeness of the stories increased with each retelling. In such an environment, it was easy to portray the problem as bigger than reported and the government response as inadequate.

These attitudes should not merely be taken as political opportunism from Afrikaners. Contemporaries frequently but incorrectly accused Smuts of using Chinese labour and anti-capitalist rhetoric merely to push for greater freedom from Britain.[53] Even Selborne thought: 'the Boers have passed as opponents of the Chinese only because they are such good party politicians... There is only one section in the Transvaal who are really opposed on principle to Chinese labour, and that is a section of the British miners'.[54] Historians have agreed. Hyam and Martin are typical in arguing that Afrikaners deliberately utilised the Chinese issue to divide and rule the British in Britain and the Transvaal but did not actually care about the issue.[55] Certainly, Botha was happy to warn the colonial government in 1906 that, if the Chinese dynamited any more farmsteads, he would send a delegation to Britain demanding the immediate repatriation of all the Chinese.[56] However, dismissing this as political opportunism fails to take account of the longstanding suspicion of the mines prevalent in the Transvaal long before the Chinese became a political issue, or the very real farmstead fears so prevalent in 1905–06. They were part of the panic, while they also sensed a political opportunity. Steyn was 'concerned... for our people in the neighbourhood of Johannesburg with all the Chinese horrors. I do not know if it might not perhaps be useful to warn the magnates and Governments seriously that if they do not stop sowing the wind they will surely reap the whirlwind'.[57]

Particularly in rural areas, it is clear that there was genuine alarm. Feelings of vulnerability were particularly strong because, after peace was declared in May 1902, all Afrikaners, and many non-whites, were forcibly disarmed as a security measure. Given the genuinely isolated location of many farms, it is not surprising that they worried about their safety, even if their fears were unreasonably directed towards the Chinese. One book on the subject had an entire chapter entitled: 'The Growth of Terrorism', which described the 'notorious cases' involving the Chinese, including 'murder, rape, robbery with violence, and that class of criminal assault with which we deal in England under the Criminal Law Amendment Act', the 1885 British Act which updated punishments for homosexuality among other things.[58]

Furthermore, while many complaints were from Afrikaners, some were from British farmers. In one case, Charles Bain even identified himself as 'one who assisted in a small way in procuring the importation of the Chinese'. This did not stop him declaring: 'I do not see why the lives of persons living here should be jeopardised for the benefit of European shareholders'. The Chinese were 'wandering all over the country at their own sweet will, and are a source of the utmost danger to the inhabitants of outlying farms and small hamlets' while 'the mines are altogether under[-]policed'. He reluctantly stated that if the mines did not start paying for greater state security, 'there will very shortly be a public outcry against the present conditions, which I shall be compelled to take part in'. His agitation was because he had been involved in 'a most horrible occurrence' on 11 July 1905, in which four Chinese men 'were seen by a kafir cooking some food down in the vlei'. When they told the Chinese that they were trespassing, the Chinese 'presumably did not understand him' and 'whipped out knives with blades measuring about 14 inches long, and made an attack upon these kafirs. – The kafirs bolted', but not before one was stabbed in the back and around the face and head.[59]

His chief concern after this incident was not for the welfare of his African workers, but for his family, who frequently were left alone on the farm overnight while he was away. Their safety clearly depended in his mind either on his own presence or on SAC men being available. He argued that the problem was the lack of SAC men and that 'they have to bear the cost of the journey themselves, which of course does not induce them to go out of their way to arrest stray Chinamen and bring them back to town'.[60] The rumour that the SAC who caught Chinese were not getting subsidised for the effort involved was false but such rumours angered British settlers and diminished faith in the ability of the government or mines, so closely aligned together in the public mind, to handle the Chinese.[61]

On 12 April 1906, a meeting in Heidelberg (scene of the only 1903 Afrikaner petition in favour of the scheme) complained 'that parties of roving Chinamen have almost daily been seen about the district, and at times murderously attacking unprotected and peaceable inhabitants, both male and female'. On the 18th, over 400 people met in Heidelberg to request 'respectfully' for the 'Government to afford better protection to its subjects'. They mentioned 'murderous outrages' and 'roving bands of Chinese', for which they held the mines and government jointly responsible.[62] C. P. Crewe, the Colonial Secretary, wrote to Selborne in September 1905 to warn him that 'the compound system' was too free and the mines failed 'to secure the public from danger sufficiently'. Because of this, many like him were turning against the mining industry, which did 'not recognise the dangers which the general community suffer from the indifferent control they exercise over the Chinese laborer'. He warned: 'A few more scares and feeling will be so strong that I am convinced an outcry will be raised and those who have here hitherto supported Chinese labour as a necessity will be amongst the strongest opponents of its continuance.'[63]

In reply, Selborne sought to dismiss such alarm 'as exaggeration... natural when you get half a dozen serious crimes, of which two are murders, following each other within a fortnight; but the fact remains that the matter has been viewed out of perspective'. There were about 45,000 Chinese men employed at that time, after all, and only 'a small percentage of ruffians'. The majority of crimes he dismissed as being due to Chinese men losing their way on the Rand; 'the stragglers get very hungry, he goes to a farm house, all he wants is food, but no one understands him and a fracas or even a crime may ensue. This was certainly the case when poor Joubert was murdered'. In order to protect whites, however, the government swiftly passed an Ordinance 'giving every white man authority to bring every Chinese he finds off the mining property to the nearest SAC post and receive his expenses', regardless of whether he was actually an indentured labourer.[64]

Jeremy Krikler has described similar hysteria among 'white rural-dwellers' in June-July 1904 about an 'impending black revolt' in the Transvaal. As in the Chinese panic, there were large Afrikaner deputations to the government requesting rearmament[65] and the government thought it was a political ruse so Afrikaners could fight again. Still, the government eventually sent a hundred SAC men and hundreds of rifles were distributed in the affected areas, such as the north.[66] As Godfrey Lagden explained, even though he thought the alarm was mostly 'manufactured', still 'wherever white men, whether they be Boers or English, are living amongst savage races... it is necessary for their prestige that they should be armed'. Krikler directly links the end of the panic with their rearmament.[67]

The 1905 scare did not end so abruptly. Eventually, rearmament was partially allowed in late 1905. And rearmament was important, a symbolic representation of the colonial man's ability to protect his home and family and workers in true paternalistic style. However, the government's offer was hardly attractive, as only a few very old shotguns were made available, which residents would have to lease from the government for three months and with a £100 deposit each, an impossible sum for most Transvaal residents to obtain. Another Afrikaner deputation complained: 'we are British Subjects who have tried to conform to all the laws... We only want to protect ourselves, and when these stipulations are made we do not consider that we ought to take them'.[68] It was not until April 1906, after the panic continued, that the Lieutenant-Governor finally admitted 'people *must* be protected and if they can't afford to buy the necessary weapons for their protection we must do so' and determined to worry about getting the expense back from the mines at a later date.[69] This still failed to entirely pacify the situation because rearmament was only one facet of the panic.

Instead, many increasingly focused their anxieties on the mines and their relationship with the government. This was especially fuelled by the refusal of the mines to compensate victims of alleged Chinese attacks, or to compensate the government for extra expenses like the shotguns. In most cases, victims depended on public subscription lists publicised in the local press; such public appeals further spread reports of 'outrages'. While privately, the colonial administration condemned the mines' stance, publicly senior administrators such as Richard Solomon, the Acting-Lieutenant-Governor, and James Jamieson, Superintendent of the FLD from mid-1905, defended the mines, perpetuating the general impression of government-mine collusion.[70]

When Solomon wrote to De Jongh, then President of the Chamber of Mines to suggest they pay compensation without admitting liability, the Chamber did not even acknowledge his letter for four months, then wrote that they would not.[71] Selborne even wrote personally to request assistance in particular cases when he was petitioned by settlers who had been attacked but 'without any satisfaction'.[72] When the mines continued to deny responsibility for attacks, Selborne complained: 'I have no hesitation in saying if these outrages continue there will be a strong agitation for the repatriation of Chinese labourers which it will be difficult to resist.'[73] Even in the high profile murder of Joubert, which happened near one of the ERPM mines, the company denied responsibility. Because of the intense public pressure, George Farrar gave £40 of his personal money to the fund for the widow, but refused to accept any official culpability.[74] While privately, he was cooperating with the government to improve mine security, this did little to affect the public discourse, cementing the impression that the mines cared little about the lives of southern Africans, but merely wanted profit, and that the government was complicit.

Even the imperial government was drawn into the local matter, and also shows how genuine the fear and resentment felt by locals could be. A British farmer, Charles Moss, wrote to Lord Elgin, Secretary of State for the Colonies, of an attack on his sister's house by 'the yellow brutes' near Boksburg in late 1905. He forwarded her letter, which explained how the Chinese broke in and tried to kill her husband and children but her husband managed to kill a Chinese man instead, a fact which she described in her letter with evident pride. 'He is the first to kill a Chinaman stone dead and has been congratulated by everyone, for of course, everyone is full of it.'[75] The gun ownership and a celebration of her husband's ability to protect his home were given leading prominence in both her letter and in newspaper coverage of the incident. This was especially contrasted with the African workers on the farm, who were described as 'really frightened of the Chinamen' and largely hiding.[76]

After all of this, F. Hellman, manager of Farrar's mine 'at which the Chinese who committed the outrage were employed', refused to compensate them, saying they could not 'pay for all the criminals in the country'. Charles Moss complained to Elgin that this exemplified the way 'mine owners shirk their responsibility... At the present time the mine owners are the real rulers and their power has been increasing in an alarming degree, so that all interests are becoming subservient to their jurisdiction'.[77] Yet Elgin, although moved by their story, found he was legally powerless to interfere in ERPM's decision. Such a stand-off only fed the ill-feeling on all sides and ensured that stories about 'outrages' perpetuated and became increasingly politicised.

Riots

Riots were another important factor underpinning the panic and directing criticism towards the mines. One tourist account told of a 1904 riot at the Van Ryn compound: 'a policeman was stabbed by a coolie, while two others had their brains battered out with jumpers; a fourth was so seriously injured that he was not expected to live, and a fifth was just saved by the timely arrival of the compound manager'. Another one on 11 December was attributed to 'a black Helen', causing a fight between the African and Chinese workers'.[78] Often, newspaper and book coverage focused on the cases without known causes, which fed into the idea of the Chinese as 'inscrutable'. For instance, The *Pretoria News* wrote about 'Chinese Undesirables at Durban. An Alarming Affray. One Man Killed.' An inspector found a dead body at the depot in Durban among Chinese awaiting repatriation, 'with his skull split open and four others were suffering from knife wounds'. Another died during the night from his knife wound. Five Chinese were arrested but 'the police have been unable to determine the cause of the fracas'. As usual in such cases, there was always a 'curious part of the affair'; in this case, it was that, when inspector entered the compound, almost all the men in it pretended to be asleep.[79]

In reality, one of the most common reasons for unrest appears to have been wage disagreements, even though the detailed legal contract was supposed to have solved the problem. Several disturbances seem to have arisen because the mines decided illegally to switch to a piece-meal system without drawing up a new contract. There were also several uprisings about whether a month of work was a calendar month or a lunar month of 28 days. The Chinese were unfamiliar with the Gregorian calendar and had assumed they would be paid by a lunar month. Likewise, the mines frequently deducted money for supplies provided upon arrival (such as oil skins and boots), for time in gaol, and for allotments sent back to their families in China. There were also widespread deductions of wages for poor work performance. What was deemed poor performance varied from mine to mine, as did the punishment exacted. The government was willing to allow pay suspension if a Chinese worker failed to work at all.[80] Few cases were so clear, and the mines implemented a policy of 'A fair day's pay for a fair day's work'. Holding back wages, and in some cases even food, was meant to eliminate lazy Chinese labourers and, the mines argued, would increase both the average pay for all who worked, and the workers' efficiency. While the Chinese Consul in Johannesburg later complained and the British and Transvaal governments tried to define 'a fair day's work', to eliminate this problem, no conclusion was reached.[81] While some of these occurrences were undoubtedly genuine misunderstandings, it is clear from the records that some mine managers utilised the contractual ambiguities to save money.[82] They had plenty of scope.

Corrupt Chinese police were often blamed for this type of unrest on the mines, conveniently absolving the mine management of any responsibility. For example, on 29 July 1907, the Cason Gold Mine had announced that the Chinese would only be paid if they drilled at least 30 inches in a day; the previous agreement was 24 inches. This new contract had been 'posted in the Compound for 15 days, the notice was printed in Chinese, and was posted in a conspicuous place in the compound'. The Chinese police were expected to make sure the workers under them understood the new terms and cooperated. Some of the police organised a meeting instead to ascertain whether support existed for the changes or not. The Compound Manager, upon hearing of the proposed meeting, called all headmen to his office and asked if they had any complaints, suspecting the meeting meant unrest. After a short discussion, the Compound Manager thought they went away 'perfectly satisfied', promising to work the next day as usual. But on Monday morning, just over half the shift appeared:

> I got a message that the coolies refused to go down and were throwing things about; I went over to the shaft and I heard the coolies beginning to shout strike him! Strike him! I found two coolies, one at each shaft head, preventing them from going down; all this time the coolies were crowding up round the entrance and one of them blew a whistle and they then surrounded me and I could do nothing. I was rescued from the mob by 5 or 6 of the Chinese police and as I was going away I saw the Coolies rushing to the dump, and picking up stones[.]

A major riot ensued and the Transvaal police had to be called. Several shots were fired; a Chinese policeman died. The accommodations of white officials were ransacked.[83] Despite the fact that Chinese police had rescued the Compound Manager and been injured, the riot was entirely blamed on the duplicity of Chinese police. In reality, this was about the limited powers Chinese police had over the rest of the workers. The Compound Manager assumed the positive feedback he received the night before was representative of the entire Chinese community, whereas the riot makes clear this was not the case. Indeed, at one point, the Mine Manager said he heard the rioters cry 'kill the [Chinese] police', possibly because they were associated with the general management decision. The mine manager, keen to shift the blame, thought 'gambling was a probable cause of the row, and the inch question was made an excuse for the riot'.[84] Indeed, the assistant mine manager thought it merely because 'the Chinese have an extremely excitable nature and I think that 3 or 4 men could easily sway the whole crowd'.[85] Another senior mining manager blamed it on an 'ex-Head Policeman who has since been repatriated and who at the time of the riot was not at the Cason compound' but 'left an extremely bad influence behind him'.[86] Another remarked 'they have calmed down so quickly this time I do not think that there was very much grievance'.[87]

Such riots demonstrates both how reliant the mines were on the Chinese police and how ill qualified they were to understand internal Chinese conflicts. One riot actually involved 'an organised attack on the quarters of the Chinese Police'.[88] The only conclusion the FLD inspector could draw was one of subversion linked to documents and insignia relating to a secret society, the 'Ko Lo Hui', found in the quarters of the attackers. How or if this was connected was never clear since no major investigation was launched. When relations between Chinese Police and Chinese workers affected only non-whites, mines neither grasped the ins and outs of that relationship nor considered it a priority. After all, not understanding the Chinese conformed well with the characteristics of inscrutability and an unwillingness to assimilate attributed to them in racial discourse. It also fit in with the stereotype of the corrupt Chinese official. One British newspaper reporter claimed to have seen a 'form of punishment, which I will call Oriental or degrading', which included 'the use of the head-board or wooden collar with a hole in it'. Such a punishment would supposedly degrade the coolie without injuring him. 'The instrument of all these practices were, of course, the Chinese police...I don't blame mine owners for it; it is probably an inalienable part of the worst type of Chinese officialism.'[89]

After one of the first large riots on the Lancaster mine in September 1904, with Chinese police leading their men, A. W. Baldwin, a white miner, demonstrated an uncharacteristic sympathy for the Chinese police's dilemma:

> These men are responsible to the villages and places from where they gathered their coolies together. Should they return to China without the coolies it would possibly result in their death. They are bound to stick up for their coolies in all trouble that arises here, and should they not do so, it might go hard with them...the coolies forced their hands and they had to accompany the men in making a demonstration.[90]

Such a position jarred both with public outrage and with the mines' desires to deny culpability. He was accused of sanctioning the Chinese police riot and the mine dismissed him.[91]

Despite racial prejudice, there was clearly corruption and excessive violence. Records of this sort of abuse are probably incomplete as cases were only brought to the attention of white supervisors when other illegal activity was involved or when a large fight occurred. In one case, all of the Chinese interpreters and a white miner were dismissed from the Lancaster Gold Mine in 1906 for running gambling and opium operations on the mine.[92] In another, six Chinese workers charged with desertion claimed that 'the Police boy beats us because we won't lend him money and so we ran away'.[93] Indeed, Chinese police were frequently cited by Chinese labourers when caught breaking regulations or committing crimes. In one case, a petition was presented to the FLD, alleging that two Chinese police had used their position to have 'illicit and unnatural intercourse with the youths of this mine...If a boy refuses to entertain the illicit suggestions of the Police, he is beaten unmercifully'.[94] While the forced sexual acts were never proved, and this particularly case was kept out of the press, accusations that the Chinese police enforced order through beating or even sexual assault were common.

Chinese police were therefore often blamed for mine problems. After a disruption on his mine, for which twenty Chinese police were convicted, J. W. H. Stubbs was moved to remark: 'These Headmen...are no benefit to the Company. They do no work, but dictate to the Compound what work must or must not be accomplished...they are a source of the very worst danger to the mine...I am sure they are a treacherous and unforgiving tribe.'[95] The men were given three months hard labour and then became the first criminals to be repatriated. It would become an increasingly common 'punishment', to simply ship 'bad characters' to China and ban them from returning. It was difficult replacing Chinese policeman, however, until late 1905 when the mines were finally granted permission to promote a normal Chinese labourer to the position of headman, to replace those hired in China who proved unsatisfactory.[96] Because the mines assumed this was part of the Chinese 'character', they ignored underlying factors, while unintentionally enforcing the public prejudice against the Chinese presence.

The Transvaal *Star* was typical in reporting that 'on some mines the coolies are the slaves of the police and headmen, who extort bribes and rule generally with an iron hand'.[97] While this was a regular part of violent 'supervisory' relationships on the mines, public reactions were quite different because the violence fed into the general discourse of violence and panic at the time, and because the very large number of Chinese desertions were blamed on such treatment, and most crimes were thought to be committed by those deserting Chinese. While desertion was not a new problem (both African and white labour had resorted to desertion and rioting on occasion)[98] an African or white deserter would simply blend into the southern African landscape. The Chinese were exotic 'others' and stood out. Many of the desertions may simply have been some Chinese losing their way on the empty veldt or to avoid mistreatment. Many of the deserters left with a pass, obtained legitimately or otherwise, and then simply did not return; others simply wandered off compounds. Security on some mines was allegedly so lax that Chinese could come and go at will. They also had ready access to the stores surrounding each mine, as the Ordinance had ensured they were not restricted to a closed compound. While they roamed, there were common misunderstandings between Chinese workers and neighbouring storekeepers, resulting in further violence. Unscrupulous sellers were even known occasionally to supply the Chinese with knives, guns, opium, and other items which, while not illegal, could then be used in attacks.[99]

Chinese-African violence

While the press focused primarily on stories which involved whites in some capacity, most mine violence were smaller-scale fights among each other or with local Africans. Various African groups had often been kept separate on mines, based on the assumption that they would fight if mingled. There were similar assumptions about the Chinese. The mines had been advised by Warnsford Loch, General Manager of an Australian mining company in the Federated Malay States, that 'there is very bitter class hatred and faction fights are quite common' among the Chinese workers.[100] Skinner had stated that Northern and Southern Chinese had poor relations so the mines 'arranged that...those from Southern China are not mingled with those from Northern China'.[101] This quickly became standard practice on the mines and also perpetuated the notion of the Chinese as clannish.

Policy towards the mixing of African and Chinese workers was more varied. Only a handful of mines ever had enough Chinese labour to fully eliminate Africans from their compounds. Consequently, in public, the mines downplayed any inclination towards violence between the Chinese and Africans. Instead, they claimed that Africans would ape Chinese habits, as they were supposedly prone to do with more advanced races. There would be some conflict, but only what was expected between less advanced races and nothing alarming.[102] On some mines, management sought to focus these assumed 'tribal' conflicts into sporting competitions between the two. Such activities were clearly attempts by the administrators to channel what they saw as a natural inclination for violence into a 'legitimate', approved sporting arena. Every summer, commentators, tourists, and postcard sellers would gather to witness such 'healthy' activity.[103] This acted as a sort of choreographed symbolic violence, allowing them to compete and express their virility and masculinity, but in a controlled environment. This could be particularly effectual when language communication was such a problem between different groups.

In reality, however, the records of relations between Africans and Chinese hint at considerable ill-feeling and disturbances on mixed mines were common. The press usually depicted any conflict in more threatening terms and used such instances to advocate segregation. One long article in September 1904, entitled the 'CHINESE AT PLAY', reiterated the common view that 'Tribal or semi-religious factions existing among rather savage men who have to meet one another daily inevitably lead to blows'. When Africans once accidentally interrupted 'the "mysteries" that have formed part of nearly all Oriental and ancient religions', the Chinese attacked, although the police intervened 'before any blood was shed'. The newspaper went on to make some stark warnings, however: 'First we were reminded that an outbreak involving a faction fight of many thousands is possible... every effort should be made to keep the Kaffirs and the Chinese apart. They are too expensive to be allowed to play at Lerthodi [sic] versus Jonathan, although it is a *natural* and diverting recreation.'[104]

The bias of the records makes it very difficult to understand exactly what the problem was. One early disturbance in December 1904 on the Witwatersrand G. M. Co. highlights both the problem of motive and the seriousness of such incidents. During a fight between local Africans and Chinese miners, one Chinese and three Africans were killed and more than thirty Chinese were injured. The Superintendent's report condemned the mine's failure to remove the local 'kaffir location', as previously recommended. While several reasons were given by the people involved, including one claim that a Chinese man tried to steal a dog or that a 'kaffir' struck a Chinese policeman, 'some 50 Chinese and about 300 kaffirs were fighting'.[105] However, because this involved no

Europeans, it received little public attention despite the scale. Interestingly, the white Assistant Compound Manager was dismissed for being drunk during the incident and thus failing to calm the situation before it became so inflamed.[106]

Another serious incident occurred about two miles from the Consolidated Langlaagte Mine in July 1905, when some Chinese went fishing by a 'native' location. The Chinese claimed a group of Africans attacked them without cause. The Africans claimed that a Chinese policeman and labourer 'made indecent overtures to a kaffir woman there, and, not getting what they wanted, they attempted to use violence'. The Africans defended the woman; the Chinese fled but 'incited' labourers at the compound to return and attack the location occupants. Having noticed 'Chinese coolies loitering about the vicinity of this location', the FLD inspector (usually the sole investigator in such matters, although sometimes the mines organised internal reviews, and sometimes the police investigated very violent cases where they were called in) decided to believe the Africans.[107] Soon after, the location was moved to prevent further interaction between the two groups.

In one of the most serious disturbances, a fight broke out between an African and Chinese man on the North Randfontein in September 1904, with the Chinese man hit 'on the head with a piece of iron'. The official report decided that the Chinese police then 'incited their coolies to attack the Kaffir. He however escaped and the labourers unanimously refused to go to work unless and until the Kaffir was handed over to them for punishment'. Because this was refused, the Chinese refused to work. Later that afternoon, 'the Chinese armed with sticks, jumpers, bars and other implements attacked a Kaffir location' near the compound. Two or three (white) Transvaal police easily subdued the Chinese, but the mine management was faced with the insubordination of their Chinese police, as well as the conflict at the 'native' location. The Chinese police were rounded up and informed they had acted illegally, but 'made the lame excuse that they had been to great trouble in collecting coolies in China and that when one of their number was attacked, they thought it their duty to stand together in his defence, and that, not knowing the laws of the country they had wished to punish the Kaffir themselves'. The police were prosecuted regardless (although no one else was), and were warned to report any quarrels in future. The mine officials promised to tell the Africans likewise.[108]

So how is one to explain such violence? Language was almost certainly a factor, with communication difficult at best. Traditionally, African labourers and white supervisors conversed using a form of Pidgin

English but almost all the Chinese arrived without any knowledge of a southern African language. Consequently, the interpreters were meant to aid in the smooth administration of the scheme. This often failed to happen. There were no regulations to govern how many interpreters had to arrive with a shipment; the superintendent of the FLD only managed to force the mines to hire European managers who could speak a little Chinese at the end of 1905. Alfred Child, the most senior Chinese Advisor to the Chamber of Mines, explained during a criminal trial that there were 'several different races of Chinese on the Rand ... I have come across one from Manchuria and one from Korea... They all speak a sort of Mandarin, which broadens as the provincial English broadens... South of the Yangste, Cantonese is spoken and another language'.[109] The formal Mandarin-speaker was often incomprehensible to the workers, but what most Europeans employed spoke, if they spoke any Chinese.

In addition, the Chinese and African labourers, most of whom were from Mozambique, were unfamiliar with the Transvaal legal system. The fact that both Africans and Chinese probably felt themselves to have no legal recourse in cases of dispute, and so took action into their own hands, probably did not help. It is also likely that many Africans had also been largely against the importation of Chinese workers, fearing they were meant to replace them.[110] While some conditions on the mines improved because of the Chinese arrival – the availability of hot water and more spacious and better ventilated accommodation,[111] for example – other aspects of African life became worse. For one thing, throughout 1905, there were increased Africans recruited but not enough jobs for them to fill, so their wages dropped. Some of the possible reasons for conflict were fantastical, such as the unsubstantiated claim that 'witch doctors had told the natives that Chinamen eat babies' or slightly less fanciful, that 'some desperate battles took place when Chinese bands attacked kaffir kraals searching for food'.[112]

Chinese racism may even have been a factor, as the Chinese seemed to have responded particularly badly when the mines periodically used their African police, the counterparts to Chinese police, to quell disturbances on the mines. By mid-1905, the Lieutenant-Governor was driven to point out that using African police to control coolies was 'liable to inflame the coolies against the Kaffirs and it is impossible for the authorities to maintain proper discipline if similar action is taken by mine managers'.[113] One manager's letter on the situation was adopted by the CMLIA and distributed to the other mines to highlight best practice. They were never to allow 'native' or Chinese police to interfere with each other in any way and pointed out 'the utmost importance to ourselves and the whole of the Mining Industry that no racial differences be created between Chinese and Natives... in fact efforts should be made to establish the best possible relations between the two races'.[114] It is difficult to know, from the records, to what extent their rational was based on actual fact, however, or how much was speculation.

Most explanations offered in investigations involved some sort of dispute over property or women. The Acting Secretary for Native Affairs insisted 'considerable irritation has been engendered among the Natives by the petty thefts of these Chinamen'.[115] The General Manager of the Geldenhuis Estate complained there had been an 'extreme nuisance' caused by the Chinese from a neighbouring mine, who

> steal anything they can lay their hand on, having stolen fruit from my own garden, and fowls, pigeons, etc from some of my employees... they frequently commit nuisances in public places, such as near married quarters, etc... if their prowling habits are not stopped immediately, there will be trouble between them and my Natives, several quarrels being stopped, yesterday, just in time to prevent serious trouble.[116]

Conflict clearly occurred for a number of reasons, although the administrative records only recorded the violent conflicts which were brought to the attention of whites, so the records are undoubtedly skewed. Likewise, far more non-violent encounters must have occurred daily, but there is no recording of these. In fact, in one case, an official complained because he thought the Chinese and Africans were getting along too well.[117] Because of this, the records give the impression of frequent, and often unexplained, violence.

Given that some historians have argued that mines often encouraged inter-ethnic strife as a way to control labour,[118] it is worth pointing out that this does not seem to be the case with the African and Chinese workers. While some competitiveness was no doubt encouraged, particularly in mine-organised games, actual conflict seriously affected mine productivity and peace. In January 1905, the Secretary of Native Affairs investigated the claim that fights over women were the possible causes in several disturbances. He recommended they 'remove all mine locations where Native women are resident from the vicinity of Chinese compounds'.[119] The mines could not afford the luxury of encouraging a conflict which they could not control. When violence occurred, segregation was the usual solution. After one disturbance, the mine involved not only moved the native location, they also discharged all the Africans employed by the company on the mine, apart from the 'boys' working on a Mill and Cyanide Works, who were then kept away from the Chinese.[120] On mines where the elimination of African labour was impossible, it became usual to have all Chinese and African miners work separate shifts in order to still ensure as much segregation as possible.[121] The results of this discourse would be to firmly enmesh a system of racial segregation within gold mining, later applied to all aspects of South African life. This was not simply a matter of control, of divide and rule, but reflected an easy way to address conflict without having to deal with the more complicated issues raised. The problem also highlighted a sense of the powerlessness of white miners and colonial officials; each case reminded them how superficial their control and their understanding of Chinese and Africans was. Thus segregation was mostly about fear, a panic rather than a Machiavellian control strategy.

White on Chinese violence – corporal punishment

Of course, this was not the only type of violence on the mines. South African whites had a long tradition of utilising beatings to control their African labourers, without international attention.[122] Likewise, South African newspapers were often sympathetic towards white men who attacked Chinese men. Indeed, the Labour Importation Amendment Ordinance No.27 of 1905 made it lawful to arrest without warrant any Chinese man found outside of mine premises, and considerable force was sometimes used.[123] For Afrikaners, this meant a chance to glorify the power of traditional commandos again, with bands of men 'assisting' the SAC to capture Chinese men, who were assumed to have nefarious intentions. This was also a chance to symbolically assert standard colonial masculine tropes of protecting their families and homesteads in the face of external threats. Newspapers and other accounts always valorised accounts of men attacking Chinese 'invaders', especially if they died in the attempt, as Joubert had done.

More problematic was the issue of violence towards the Chinese on the mines themselves. Initial attitudes were sanguine. The matter became controversial when a particularly brutal assault was brought to light by a reporter for the *Daily News* in Britain in March 1905, just prior to Evan's departure from the FLD. In the case reported in the press, a Chinese deserter was captured who then had to be admitted to hospital for twenty days in order to recover from a beating he had received.[124] The newspaper report led to an investigation into the reasons for desertion. Two cases 'of a particularly disgraceful and brutal character came to light', which led Evans to disclose to his Secretary (and his temporary replacement), Wolfe-Murrey, that he had indeed 'given his sanction to the infliction of slight corporal punishment for minor offences on the Mine',[125] but this went beyond what he had envisioned. 'I will not have men wandering about complaining of flogging and having such noticeable scars to shew.'[126] Not only were the revelations damaging in Britain and China, this incident fed a growing belief in the Transvaal that most crimes were committed by Chinese deserters who sought to escape from white abuse. The fact that many deserters blamed harsh treatment for their actions ensured that the violence on the mines seemed important to Transvaalers; they could not ignore the situation in the way they could ignore violence towards Africans. This contrasted sharply with earlier indifference to such abuses, and perhaps also explains why concerns over Chinese 'slavery' did not translate into general concern for African labour conditions.

When the government finally cancelled the use of corporal punishment in June 1905, it was explained that 'this permission to inflict slight corporal punishment has been abused both by the white mine officials and the Chinese police', which in turn led to high desertion rates.[127] The judge, Rose Innes, claimed through his experience overseeing criminal cases against the Chinese that the conditions of indenture 'caused them to desert and terrorise the countryside'.[128] Likewise, James Jamieson, the newly appointed FLD Superintendent of Chinese, blamed most Chinese violence on white supervisors being violent: it was understandable retaliation, further encouraged by the unwillingness of magistrates or juries to convict white men of abuse: 'I am absolutely convinced that a Chinese rarely has a chance against a white man in Court, and justice ... has been in the nature of mockery, and realised to be such by the coolies.'[129]

Conversely, many miners blamed the 'outrages' on the government interfering in their ability to use corporal punishment. They pointed to the fact that the practice was banned in June 1905, followed by months of increased attacks. The problems occurred because the mines were left with no means to effectively discipline the Chinese. After a disturbance in August 1905, for instance, one general manager complained: 'as long as the control allowed by the Companies over the Chinese is so restricted these disturbances will occur and will increase in frequency ... unless lashing is prescribed for desertion and loafing this prospect appears to be very serious'.[130]

Increasingly, however, senior mine managers were willing to fire whites whose violence was linked to unrest. One case aptly demonstrates this:

> One of the Compound boys was returning from Boarding House with his 'boss's' breakfast. On passing one of the locomotive drivers the white man gave the pannikin a kick. The Chinaman afterwards told me that he was so infuriated to see his master's food so treated, that he called the white man a 'b....y fool' an epithet all the Coolies have learned and frequently apply, not knowing its significance. The white man thereupon picked up a stone and struck the Chinaman on the head, knocking him almost senseless. The news soon reached the Compound that a Chinaman had been maliciously killed by a white man. The Chinese Police rushed down, found the white man, and attacked him. Some one then ran for the White Police. By this time the Coolies were so excited and poured out of the Compound, getting sticks, jumpers and rocks for weapons... I may mention that we immediately discharged the offending white man, as I consider that he was responsible for the whole trouble.[131]

Once banned, cases of physical abuse did not disappear, of course. Miners continued to utilise violence, sometimes for the flimsiest of reasons. On board ships it too was often used, and with even less regulation than on the mines.[132] Accounts from white miners themselves contrasted with public fears; they often portrayed violence as a necessary, even prosaic, aspect of the indenture scheme. This banal attitude to violence can be seen in a letter written by an American miner to his mother, describing a mine fight:

> a Jew came running across the veldt, followed by a lot of Chinamen from the Durban-Roodepoort mine and took refuge in the boiler house of the West shaft... the door opened and in came the Jew, followed by Chinamen, sticks, stones, pick handles, and jumpers... Stockett at once called up the Kafirs from the compound and a small battle was averted only by the Chinamen handing over the Jew to Stockett. The police had been telephoned for, but Dr. Lacey, Manager of the Durban-Roodepoort, got there first and took the law into his own hands. Jumping from his horse, he made for the nearest Chinaman, relieved him of his pick handle and then proceeded to do him up. He cut his face open with his sjambok and gave him a most unmerciful thrashing. It was funny to see how quickly the rest

of the Chinamen dropped their weapons. It is wonderful what one determined man can do with a mob... The only other exciting thing that happened this week, was a dog fight... Oh yes, a kafir was killed in the shaft. [133]

Describing the brutal suppression of the Chinese attack, a dog fight, and a mine accident which resulted in an African death in such a nonchalant fashion was easy in an environment where life was held so cheaply and violence so commonplace.[134] While violence did not stop on the mines, there does appear to have been a small increase in the number of white workers then prosecuted for the violence, and the government launched several investigations into the issue. Violence largely remained 'legitimate', however, as long as it was not so severe as to cause disruption to the mines.

Sex

There was not merely an obsession with violence, however. One of the biggest obsessions was Chinese sexuality. While many fretted about the alleged Chinese propensity for 'unnatural acts' – that is, homosexuality – others were obsessed with the idea of the Chinese breeding a new 'bastard race' with prostitutes in the Johannesburg area. Homosexuality was thought to be widespread among Chinese men, particularly Chinese migrants, who were almost wholly male. The Chinese were also thought to be sexually rapacious, as the population of China was so large, so much commentary on the experiment focused on the possible outlets for Chinese sexual need. The mines were required to promote the importation of women but traditionally Chinese women were made impure if they left Chinese soil. Nor did the mines provide married accommodation for the Chinese, although legally they were required to do so. About thirteen wives accompanied or later joined their husbands to the Rand, and about 25 children, mainly boys, although most wives returned soon after arriving, due to the lack of suitable facilities for them.[135]

In general, the mines were either unconcerned or unwilling to address the sexual appetites of the Chinese. Critics of the scheme, however, saw this as one of its greatest problems and it fed the panic. Salacious, titillating yet vague accounts filled newspapers, telling of Chinese 'too ill to work, on account of their excesses'.[136] Indeed, even the relatively neutral James Rose Innes, the Chief Justice in the Transvaal, went so far as to blame the lack of women as the chief reason for desertion from the mines. 'To confine thousands of celibate labourers in compounds, from which a considerable number were bound to escape, and, escaping, were driven to wander as vagabonds through the countryside, was a course conducive to vice and violence.'[137] Bizarrely, the belief in the sexual practices of the Chinese also led some of the ships leased by the CMLIA to secretly employ male prostitutes on ships from Northern China, and, in later years, rumour circulated that Japanese prostitutes were also imported for the Chinese labourers, although there was no reported increase in the local Japanese population.[138] The mines were willing to ignore the sexual practices of their workers, as long as it did not affect their performance. It could even bond fellow workers together, as Patrick Harries has demonstrated was the case between Mozambique miners at the time.[139]

To the government and public, however, such inaction (the male prostitutes were never made public knowledge) was simply immoral. Government officials were particularly concerned with the interaction between Chinese men and the women on the Rand. It was thought that Africans might object to the Chinese 'stealing' their women, yet also felt that, racially, Africans were the only natural sexual partners for the thousands of Chinese men. The Department of Native Affairs had regular reports sent by local officials, in order to continuously monitor the situation. The FLD was repeatedly asked to 'assure the Sect. of State that no real difficulty has as yet arisen from Chinese labourers going after kaffir women'.[140] Instead, according to Evans, the Chinese 'pay their attentions to the loose women of the nearest location and do not interfere with the respectable kaffir women. I understand that the kaffir women raise no objection to the Chinese'.[141] Despite the Secretary of Native Affairs having access to exactly the same information, and despite the fact that he was unable to prove a single case of unrest caused by Chinese men associating with African women, he felt 'serious trouble will arise unless disciplinary measures are adopted to prevent the contact of Chinese with Native women I have no doubt... It would be well to remove all mine locations where Native women are resident from the vicinity of Chinese compounds'[142] – a policy which was initiated but never fully carried out. Indeed, despite the absence of any specific cases, the Department of Native Affairs required monthly reports for its district officials detailing any problems throughout the scheme. All of the reports declared that no cases were known about in their district, but rumours suggested it was a serious problem elsewhere![143]

While there were concerns over relations with African women, the thought of Chinese men with white women caused so much concern that the FLD started a file named 'Connection with white women' in 1906.[144] Despite the lurid title, the file remained small, mostly a few pages from newspapers.[145] One case involved a Chinese man from New

Kleinfontein having willing sex with a white woman in 1907. Despite its consensual nature, the Chinese man was sentenced to 12 months and 10 lashes under Sec 19 (2) Ordinance 46 1903, which banned 'any *native* having or attempting to have unlawful carnal connection with a white woman.' [146] The administrators classified the Chinese as 'natives' without any legislature introduced to confirm this disputable fact. Furthermore, while the press often linked this incident to the indentured Chinese, the man convicted was a 'free' settler.

There were only two known cases for sexual assault involving the indentured Chinese; one involved a Chinese police man from the Consolidated Langlaagte Mine, sentenced to 6 months and 15 lashes for attempted rape on 23 September 1905. Chiefly because of the gender, race and class of the 'victim' (she was Cape Coloured and described by the court interpreter as 'shameless'), the FLD actually thought he was innocent.[147] In the other case of attempted rape, again doubts were cast on the allegations, because her family was described as immoral and, just as bad, of mixed-race.[148] The medical doctor stated:

> I examined Mrs Roos on the morning of the 8th July at 3:30 am. I could find nothing in her condition that led me to think that she had been outraged as she described. She had no injuries or bruises on her person and was suffering from no symptoms of undue excitement or prostration such as one might expect under the circumstances.[149]

In the end, no prosecution was made, despite widespread investigations.[150]

One major difference between most Black Perils and the yellow peril panic in 1905 and 1906 is the different ways the alleged female victims were portrayed. In Southern Rhodesia, for instance, prosecutors 'did not attack the character of the victim or question her sexual history or status'.[151] There was also a marked 'willingness of a jury to convict a man on flimsy evidence'.[152] Yet the cases of Mrs. Roos and the unnamed 'Cape Coloured' girl were overshadowed by prejudices against the Chinese and assumptions about lower class and 'native' female sexuality. In both cases, the characters of the two women were severely criticised and negatively depicted as sexually experienced. The fact that the Chinese man charged initially with the sexual assault of Mrs. Roos was eventually not prosecuted reflects the degree of difference between such cases. This does not reflect less public outrage, but it does reveal the strength of prejudice against poor women around the mines.

Despite this, public reactions remained intense, and panic was further inflamed when a white woman was raped and the initial Inspector's report suggested some Chinese and Africans might actually have done it together. Later, the Commissioner of Police dismissed the idea that any Chinese were involved as it did 'not bear any resemblance to outrages known to have been committed by these people'.[153] The press still continued to report a link although no one was prosecuted. Similarly, the Chief Magistrate, Johannesburg, had to deny that any 'Chinaman has ever been charge in this Court with rape, attempted rape or indecent assault',[154] despite local press reports that one had in early 1906. The public simply refused to be convinced that the Chinese were not a sexual threat. Indeed, any contact between women and the Chinese was regarded with suspicion. However innocent, sexual overtones were read into such encounters.

This was evident in 1906 when a Johannesburg photographer was found to specialise in taking pictures of Chinese workers with his female assistants (white or Coloured and described as very pale). It was apparently a common pastime for the Chinese to dress up and pay for such photographs, often with their arms around the girls (see images 3 and 4). These photographs are perhaps some of the most curious archival records of the entire scheme.[155] These images only survive because, in 1906, rumours that the place was being used as a brothel led police to raid the premises. While no prostitution was proved, the photographs were seized and the place shut down by police.[156] To southern Africans, these pictures were disturbingly sexualised, although the extensive investigation found no evidence of actual sexual activities. It is also unclear why the Chinese had the photographs taken. While there is undoubtedly a sexual overtone, with the often proprietary way the men grip the women, it is impossible to know if all the images looked like the ones which were subsequently published in the press, or if the press merely published the most sensational, to match the story. There were apparently dozens of these images found when the photography studio was raided. Yet only the ones reproduced in the press have survived. It is worth noting that the Chinese miners in other postcards and photographs were rather healthy and attractive-looking. Yet, with dozens to choose from after the photography store raid, it seems likely that the local press deliberately selected images for publication which would cause the most revulsion to their viewers. Some of the men had terrible teeth and overlong faces, while others look smug and smile. In no other image of the Chinese in southern Africa were they smiling. The Chinese character was supposed to be cold, impenetrable. And in the commercial images sold on the Rand, they were depicted as the stereotype would dictate. But in these, the smiling Chinese seem to leer, to delight in their power over white women. This clearly subverted colonial images at the time, just as it subverted colonial sexual desires. To quote Saloni Mathur, who has written about colonial postcard collecting, these usually represented a 'highly sexualized model of colonial relationships that posits the western male photographer as spectator/violator and the non-western woman as the object of his gaze/penetration'.[157] While the photographer was white and male, these were illicit pictures made not for public consumption in Europe or its colonies, but specifically for the Chinese men in the images. This subversion was clearly part of why these images were so disturbing for a Western audience and perhaps was part of the fun for the Chinese men. Faced with such images, and the sexual relations they supposedly reflected, they took on a symbolic importance; sexual assaults 'were perceived not just as an attack upon the body of a woman but as an attack upon the white community itself'.[158] Such an interpretation can clearly be seen in the reactions to rumours and the photographs of the Chinese men with women.

Furthermore, throughout 1905 and 1906, newspapers increasingly mentioned 'unnatural vice' also being prevalent and, according to some reports, were even 'teaching' Africans the 'Oriental custom'. The matter was also serious because some Chinese claimed homosexual assaults as their reason for desertion or other crimes. During the investigation of *Rex vs. 8 Chinese coolies* charged with housebreaking at Leukop on 9 November 1905, for instance, one Chinese coolie (who later turned King's evidence) said his reason for deserting 'was that the other coolies in the compound were always attempting to commit sodomy upon him, and made his life a misery to him'.[159]

Consequently, the government decided to launch an investigation to 'uncover "hidden" cases'.[160] In 1906, the newly elected Liberal government launched a secret government investigation, overseen by John Bucknill, the Commissioner of Patents, to investigated whether: 'European women are accustomed to receive Chinese indentured labourers for the purpose of illicit intercourse', the 'prevalence of unnatural vice among Chinese indentured labourers', and whether the Chinese were 'corrupting' African men. There were two specific charges which convinced the government to launch the secret investigation. Allegations were made by a Reverend Alexander Francis, forwarded to Elgin from F. Mackarness of the *Daily News*, and in a letter from Leopold Luyt, a brief visitor to the Transvaal, also sent to Elgin. Francis was a Minister of the American Embassy Church, St. Petersburg, on a visit to South Africa.[161] Luyt was, reportedly, a new arrival in Johannesburg, chiefly known around the 'the training stables at Turffontein and the various racecourses... and mixed up with a very rough crew'.[162] Another claimed he 'was a man openly keeping a prostitute, and he had so little regard for himself that he took them to a public hotel and associated openly with them'.[163] One policeman had discussed politics with Luyt once or twice and 'he is very pro-Boer and anti-English in his views, very bitter over this late war, and over this Chinese labour question'.[164] Before various testimonies sought to discredit Luyt, however, Selborne instructed Solomon to investigate the two allegations. The news soon leaked, how is unclear, but probably because of the extensive list of people questioned on the matter: the FLD Superintendent, Inspector General of the South African Constabulary (SAC), Commissioner of Police, the mayors of Johannesburg, Germiston, Boksburg, Roodepoort-Maraisburg, Springs, Krugersdorp, the president of the Chamber of Mines, Amos Burnet, principle Wesleyan minister on the Rand, N. Audley Ross, Presbyterian minister and chairman of Witwatersrand Church Council ('composed of Presbyterian, Wesleyan, Congregational, and Baptist churches'), the archdeacon of Johannesburg, and mine managers and supervisors from three mines named in the allegations. They testified 'that the crime of sodomy is not unknown among the Chinese labourers, although the prevalence of it is grossly exaggerated' although most were also not in 'a position to give first-hand evidence' either way. A few resented 'a charge which implies that the members of the white community would tolerate the existence of such a scandal without interference'.[165] The report found that, 'although in some few cases in the past have occurred in which European prostitutes of the lowest type have received such Chinese, such occurrences are extremely rare and are suppressed by the Police with a firm hand'. The report also concluded:

> That it is practiced at all is no doubt due to the different standard of morality which prevails among Western nations from that which prevails among certain Eastern races... Any practice amongst kaffirs [was] due to the fact that some of the Portuguese tribes have the Eastern, rather than the Western, view of this matter. There is no evidence whatever that they have been contaminated by the Chinese.[166]

Indeed, one 'native' minister claimed syphilis was prevalent among natives anyway: 'In places it is ghastly.' He had never personally come across a case involving the Chinese, nor had the man in charge of the Native Mission on Rand, nor had any of his African catechists or teachers 'ever mentioned the matter' which they would have done if it was common.[167] A mayor thought 'South African natives... have little or nothing left to learn in the direction of vice[;] their pernicious arrangement of plurality of wives, their vile practice of "hlabonga", their bestiality habits are proof sufficient of this'.[168] The one town official who did think there was truth in the allegations, although entirely based on hearsay, was also the only response censored from the final report.[169] The one point of agreement was that there was an increase in the number of African women in the area, there to 'service' the Chinese.[170]

As for the frequency of Chinese relations with white women, the Report found: 'although some few cases have in the past occurred (to which similar isolated cases may possibly still occur in the future) in which abandoned European or substantially European prostitutes of the lowest type have received Chinese for the purposes of immorality, such occurrences are extremely rare'.[171] Two brothels had been raided in Johannesburg in June, but the only Chinese prosecuted were un-indentured, despite local press claims to the contrary.[172] Other cases involved two white prostitutes who had 'received somewhat rough treatment from a party of European miners (who had been in the habit of consorting with them) owing to the fact that these miners were under the impression that the two women had prostituted themselves to Chinamen'. The Johannesburg photographer was also mentioned.[173]

The government was clearly unwilling to encourage what they thought were unfounded accusations of widespread 'vice' and was able to shut down public debate largely by withholding the minutes of evidence. Reporters, like F. Mackarness at the *Daily News*, chief of those who had cried sexual depravity, were told 'that quotation from them will be regarded as a breach of confidence'.[174] The only major outcome of the report was the prohibition of all Chinese theatres on the mines.[175] This was part of a move after the investigation to eliminate homosexuality. Selborne pushed forward Bucknill's recommendations that 'those officers who said that they could pick out these [homosexual] persons with a view to their repatriation should do so'.[176] The FLD circulated a memorandum to all mines and FLD inspectors with directions about how to 'detect' homosexuals.[177] Even before the findings were complete, there was a general feeling that 'gamblers, play actors', and the like were 'dangerous to the control of labourers on the Mines'.[178] Another consequence was an investigation into African sexual practices, which concluded it had been imported from Southern Mozambique, probably instigated by the Catholic Portuguese.[179] Typically, this report received almost no public attention, unlike Bucknill's, as it no longer fit into the narrative of the panic sweeping the Transvaal. Both, however, were attempts by highly educated officials 'to subsume' African and Chinese 'sex acts within Western patterns of male identification', as Forman has described it.[180] Banning Chinese theatre was part of this attempt.

Coverage vs. reality

The newspaper coverage of illegal sexual encounters was clearly overstated. The reactions to violence were similarly disproportionate to the frequency of 'outrages'. Because of the public controversy, the FLD began keeping meticulous records of all cases involving Chinese and 'natives' sent before a court so they could compare them. Statistically, the Chinese were proportionally found guilty of more crimes than 'natives', but as a proportion of overall crimes, the numbers were small. Between 1 July 1904 (when the Chinese first arrived) and 30 June 1905, there were approximately 68,000 Africans on the mines while the number of Chinese rose to 39,352. During that first year, 'aboriginal natives'[181] were convicted before a judge and jury or the Witwatersrand High Court in Johannesburg of 267 offences. For Chinese over the same period, there were three convictions, none of which went before the High Court. Crimes seen before judges or appealed to the High Courts would have been the most serious, the ones receiving the most coverage: murder, attempted murder, and housebreaking being the most common; none of the Chinese were charged with these. There were also 28 separate incidents classified as 'riots' involving the Chinese reported to the FLD.[182] Newspapers and official records suspected Chinese men in a few serious cases, but no charges were brought.

The statistics rose only slightly for the 1 July 1905–30 June 1906 period, although this covered the period of the panic and most intense press coverage of 'Chinese outrages'. In that year, there were approximately 90,000 'natives and coloureds' and 45,000 Chinese. 201 'aboriginal natives' were convicted of crimes, seventeen of which were for the murder of 'white persons'. There were 158 Chinese convictions, the vast majority, at 79, for housebreaking, nine of which were 'on white people'. Twenty-nine were for murdering 'white people'. Twelve 'natives' were convicted of rape, with seven convicted of 'assault to rape'; none of the Chinese convictions were for crimes of a sexual nature.[183]

While it is true that proportionally crime was higher among Chinese than among 'natives', these were a small fraction of overall crime; 'native' crimes, in comparison, received very little attention from the press, mines, or government officials. Partly this discrepancy was because the statistics under-report the majority of Chinese 'crimes'.

Most were tried before FLD inspectors for breeches of their contract, for which they could receive prison sentences or be repatriated. These offences would include desertion, incitement to riot, refusal to work, gambling and various other offences committed on the mine compounds themselves. In total then, there were 13,522 convictions of Chinese in that year, with 11,753 being for some form of breech of their contracts as stipulated in Ordinance 17 (1904) and the updated Ordinance 27 (1905).[184] Given most of these were some form of the Chinese being out of the compounds without permission, these statistics hardly undermined the idea that the Chinese were free to do what they wanted without sufficient controls. Furthermore, in many cases, the FLD would simply repatriate suspected criminals, rather than take them to the high court, further skewing available data.

There was also a widespread belief that Chinese labourers were behind a great many more unreported or unsolved crimes, just as the perception persisted that the Chinese were engaged in sexual 'vice'. Rumour and inaccurate newspaper reporting fed such perceptions; that is part of the nature of a moral panic. In mid-May 1906, for instance, several morning newspapers ran with the story that a white woman had been murdered by Chinese men at Natal Spruit. The newspapers never issued a correction when it turned out that the Chinese had gone to the house, but no damage was done and no white woman injured.[185] In another case, some men broke into a miner's house while the man was absent, his wife and children barely escaping alive. While no culprits had actually been seen, because 'the Transvaal has been terrorized by Chinamen, who have deserted from the mines', the Chinese were blamed.[186]

Those who realized such reports were inaccurate still took an interest in the matter. A popular Cape play throughout 1906, called *die brandende kaers* (*The Burning Candle*) was a farce about a rural family constantly in fear of Chinese invaders, hiding under the table any time there was a knock at the door – the Chinese never arrived. Written and performed in colloquial Afrikaans, 'primarily the language of the ignorant, the poorer white, and the brown people',[187] the play made fun of Transvaal fears and the lurid press reports, but also showed how widespread the stories had become, how even rural communities hundreds of miles from the Rand could share in the imagery of Chinese 'outrages'.

Despite the comic potential, the issue was an emotive one, especially among the rural, mostly Afrikaner and African, population in the Transvaal.[188] Some newspapers sympathetic to the mines, such as the Johannesburg *Star*, tried to argue that fears were exaggerated but this had little effect on halting the salacious stories.[189] Even Rose Innes, a judge likely to have a clearer idea of the actual levels of violence, thought their presence 'added greatly to the work of the courts'.[190] Perhaps with a touch of irony, statistically, the Chinese were far more frequently linked to violence as victims, although this featured little in the Transvaal panic. According to the government's own records, the Chinese were about six times more likely to die as a result of murder, manslaughter, or suicide, than 'natives' were, although the overall death rate among Chinese for much of the period was about half that of 'natives'.[191]

Conclusion

It is difficult to give precise dates for the Chinese panic. Fear increased gradually throughout mid-1905, but blew up dramatically in August 1905 with the murder of Joubert and was most intense from then until the end of the year.[192] Certain events in 1906 also spread panics, like the published accusations which led to the Bucknill Report, or the Zulu Rebellion, when police numbers in the Transvaal were reduced and concerns about 'native' and Chinese rebellion alike were heightened.[193]

The Liberal Party victory in early 1906 and their subsequent freeze on Chinese importation slowed the panic down and very few cases were reported after the scheme was entirely cancelled in 1907 as the first act of the Transvaal's elected legislature. There were moments of increased panic, like when, on the same day H. J. Walker, his wife and young sons were attacked by Chinese men from the nearby New Comet mine and a man was also murdered in his house near Benoni. An armed mob gathered at the scene quickly and rounded up any Chinese nearby and charged them with the murder.[194] These flair ups became increasingly rare, although the sensational and fearful coverage of Chinese 'outrages' did not die down completely until the final Chinese left in 1910. In fact, there was a final murder charge against a Chinese worker dropped so that he could be repatriated on the last ship; by then, justice seemed less important that getting rid of the lot.[195]

The prevalence of such violence became the most enduring image of the scheme. Part of the focus was undoubtedly fed by the stereotype of the Chinese as 'bad characters' fostered by failings in early recruitment and by subsequent riots on the mines. It is important to remember that there were actual acts of violence and sexual activity. And both the acts of violence and public panic became symbolic of wider issues of power. McCulloch has claimed that 'murders, no matter how brutal or senseless, never provoked the same degree of anxiety'[196] as rapes in Southern Rhodesia. This was not the case in the Transvaal, as the violence of Chinese assaults became symbolic of wider settler anxieties and fears. Settlers were outnumbered, sometimes out-weaponed, and often felt the colonial and imperial governments were unsympathetic to their needs. Likewise, performing violence against the Chinese was a way to show the Chinese and the government the power they held, or wished to hold, regardless of the law, while the perceived inaction of the government and mines symbolised their indifference to the safety of 'real' South Africans. The panic clearly symbolised class anxiety about 'poor whites', especially about their sex lives and 'uncivilised' violence. Even the very act of sex could become symbolic in such an environment. Chinese men having sexual relations with white women or white African men, both under white male protection, could be seen as 'an act of insurrection'.[197] While the motives of the Chinese remain the most opaque, violence was one of the few avenues available to them to assert power over others.[198] Whatever their reasoning, though, sex and violence dominated the scheme in a way which was never planned, but did have growing political ramifications for the history of the region and the empire itself.

133

Image 1 'Twixt the Jew and the Chow', reprinted from *The Owl*, 26 June 1903, p.19, Courtesy of the British Library
Caption: 'Goldbug: "I want my Chow Chow." Native: "Tens of thousands of us are waiting to be asked to accept fair treatment." Exeter Hall: "My dear friend, I might save you from slavery."'

Image 2 Postcard of Chinese labourers, reprinted in L. V. Praagh (ed.), *The Transvaal and Its Mines* (London, 1906), p.533, Courtesy of the British Library

THE CELESTIAL AND HIS BRIDE.

Image 3 'The Celestial and his bride', reprinted from William Charles Scully, *The ridge of white waters ("Witwatersrand"): or, impressions of a visit to Johannesburg, Delagoa Bay, and the Low Country*, p.215, Courtesy of the British Library

Image 4 Picture reprinted from William Charles Scully, *The ridge of white waters ("Witwatersrand"): or, impressions of a visit to Johannesburg, Delagoa Bay, and the Low Country*, p.219, Courtesy of the British Library

Image 5 'A Johannesburg Opium Den', reprinted from the *Cape Argus* newspaper, Christmas Special, 1906, Courtesy of the British Library

Image 6 'Chinese Labour', British Conservative Party election poster, c.1905, from Coll Misc/0519/16, Courtesy of the Library of the London School of Economics and Political Science

Image 7 'The War's Result: Chinese Slavery', British Liberal Party election poster, c.1905, from Coll Misc/0519/51, Courtesy of the Library of the London School of Economics and Political Science

Image 8 'Chinese or Separation', reprinted from the *Pretoria News*, 13 January 1906, Courtesy of the British Library
Caption: 'Mr. J. B. Robinson, in a recent speech, said that the new Battle-cry of the Randlords is "Chinese or Separation!"'

6
Adapting the Stereotype: Race and Administrative Control

It was originally hoped to make this chapter about what life was really like for the Chinese, to contrast with the hysteria described in previous chapter. The records make this impossible to do with any accuracy.[1] Instead, by examining the official records available on the adaptations made during the scheme, this chapter highlights 'the way the figure of the subaltern ... is both produced and then silenced by others' speech and writing'.[2] What is present in the archives, and has been discussed in previous chapters, reflects both the production and adaptation of stereotypes about the Chinese 'character' and the ways that colonials and mining companies 'silenced' the Chinese, interpreting Chinese actions and motives for their own purposes and suffused with their own race and class prejudices.[3] As the previous chapter made clear, much of the coverage of the scheme was about control, or lack of it. But when administrators were faced with real Chinese men, the stereotypes caused problems. How can you wield power over a stereotype? Analysing the ways the government and mines sought to adapt the administration to address this enables a richer understanding of colonial power and race, avoiding a dichotomous understanding of power and resistance.[4] This provides a much fuller sense of the ways power was deployed, who had power in which situations, and how that has affected the discourse within the records left behind. At the same time, it is important to understand that the increasing government interference in the world of mining presaged the tightening of controls over African labour and fostered a legislative mindset which was conducive to ideas of segregation and increased state control.

Before their arrival and the ensuing panic, the Chinese were defined by the government and mines alike as 'working machine[s]'.[5] That racial stereotype was the whole reason for their importation but this also 'undermined their ability to rule'.[6] Such a stereotypical view had been easy to maintain initially because the Transvaal had no historic experience with Chinese indentured labour and few 'free' Chinese. The mines and government also lacked a Chinese 'expert' who was also familiar with Rand mining. While there were many networks of knowledge, random public opinions or rumours about Chinese behaviour could take on a special significance in the Transvaal, which made the continued efforts to define and control the Chinese problematic. Even in the most controlled of colonies, there were always problems with knowledge accumulation, but this was exacerbated in the Transvaal. All attempts to deal with the public outrages assumed various class and race prejudices, as had the initial set-up of the scheme. Because attempts to adapt the scheme continued to be predicated on racial assumptions, they did indeed 'undermine their ability to rule'.[7]

Upon leaving office in January 1905, Milner confidently declared: 'The Labour Importation Experiment is a complete and proved success... we have... saved this country from a first-class financial smash, which would have not been without its effects even in England.'[8] He was wrong. What Rose Innes called the 'inevitable increase of crime'[9] as a result of the Chinese presence, and mine refusals to take financial responsibility for the situation, led to a growing desire for more stringent control measures. The mines and government expended far more time and money than had ever been contemplated trying to 'fix' the 'problem', and consequently, they too became obsessed with every aspect of Chinese lives. This set in motion a battle over defining the 'Chinese character', which accompanied growing conflicts over how to control them (and who would pay for it).

This conflict was complicated by the number of bodies involved, by the distance between them and by their varying interests. The implementation and smooth-running of the scheme depended upon the continued cooperation of these bodies. Their relationship deteriorated sharply from early 1905 when Milner resigned as High Commissioner and was replaced by the second Earl of Selborne. Milner had often used his considerable skills and high position to negotiate between the British government and the mines. Selborne did not command the local respect that Milner had done, nor was he willing to ally himself so closely with the mines. When the British government shifted in 1906 from the Unionists to the Liberals, largely on the back of Chinese slavery allegations,[10] he was further expected to distance government policy from cooperation with the mines while exercising greater 'independent' scrutiny of mine management. This was an environment which encouraged conflict among the interested bodies. The subsequent public and government interest in the administration of the experiment was far greater than that shown for the administration of other forms of labour.

Understanding the Chinese character and the archives

It is worth remembering that the records were only ever interested in Chinese behaviour which was considered violent or immoral or both; other aspects of Chinese lives on the Rand went largely unrecorded. Occasionally the largely white recorders reveal other information, such as the fact that many Chinese men collected clocks,[11] or photographs of themselves with European women, as detailed in the previous chapter, but Chinese motives remain obscure, further clouded by the clearly political motives of both revelations. Records of the Chinese photographs with local Johannesburg women only survived because of the panic over Chinese relations with white women. The fact that the Chinese collected clocks was only mentioned in order to assure a white readership that the Chinese might have a civilising influence on Africans.

Instead, most internal record-keeping related to the moral panic and were predicated on racial stereotyping, hardly surprising given the history of the scheme. The racialised aspects of discourse extended to who was consulted about the Chinese 'character'. In official investigations, Chinese testimony was rarely recorded. Instead, white mine and FLD supervisors explained what they were told by the Chinese and offered their interpretation of whether such information was reliable or not.[12] While the Chinese Consul might have been an obvious source of knowledge about the Chinese, the mines and government officials never consulted him. During the one occasion when he tried to offer advice after unrest, it was made clear to him that his advice was considered unwelcome interference.[13] Instead, his involvement tended to consist of bland public reassurances that the Chinese were 'perfectly satisfied with their treatment' and less violent than other groups.[14]

Instead, most 'expert' opinion came from FLD and mine figures, although they never developed a uniform view of the Chinese 'character'. The FLD staff were largely from the FO or the British Army and had mainly spent time in China itself. Its staff favoured a generally liberal interpretation of the ways the Chinese and British should interact, deploring violence towards the Chinese, but for heavily racialised reasons. The mines had a few staff with experience in China, but most white miners simply wanted to treat the Chinese as 'natives' were treated and resented any interference with this. Such different racialised perspectives often led to conflict among the different administrative bodies. In an official report about a riot at the Van Ryn Mine in December 1904, when a group of Chinese men attacked their white supervisor, Evans took the side of the Chinese. The white miner, Kennedy, claimed he refused to lend his measure stick to a Chinese worker so the Chinese worker threatened him and swore. He and a friend then 'kicked and struck' the Chinese worker, as they would likely have done with any African workers. The Chinese, however, then retaliated, with Kennedy receiving a black eye. Although the mine manager took the side of Kennedy, Evans concluded that Kennedy was an unreliable witness 'but even if true', he should 'not take the law into his own somewhat brutal hands'. Six days after the attack, Evans saw four of the Chinese involved, all of whom 'were so badly assaulted that they had to be taken to Hospital for treatment one having a broken arm'. The Chinese told Evans, through an interpreter, that 'they have been continually knocked about by their "bosses" when underground' so had 'called some of their friends together' and attacked Kennedy, but the following day they were attacked in turn 'by a number of white men and kaffirs'.[15] Despite protests from mine management, Evans wrote to the General Manager of the company to order that the three white men involved in the assault be dismissed, an unusual demand by the government at the time.[16] The manager's excuse was that the men were Australian; the mine could not be blamed for a prejudice supposedly common among Australians. Evans wrote a scathing reply, threatening to personally take the white miners to court if the management failed to punish them.[17] While the General Manager did eventually dismiss Kennedy, the fate of the other two men remained unspecified.[18]

Such incidents fed into wider disagreements about corporal punishment. The original sanctioning of corporal punishment was justified with highly racialised portrayals of the Chinese 'character'. They were supposedly imbued with a fatalistic view of life: death meant nothing to them. As part of their nihilistic approach to life, the Chinese were thought to accept most punishments as their fate and to have little regard for truth. A Transvaal magistrate provided a typical view when he described 'the fatalistic propensities of these men': 'Death to them was nothing... As might be expected, their ideas of Western justice were utterly vague; in fact it was virtually impossible to administer our form of justice among them. Like most Easterners their ideas of the sanctity of the oath were not ours.'[19] Corporal punishment was justified by insisting that exceptionally tough penalties were needed to enforce discipline and honesty as a result of this attitude. Furthermore, in China corporal punishment was common. Even the 1902 *Encyclopaedia Britannica* had stated that their 'autocratic', 'corrupt' society 'set little or no value upon truth, and thus some slight excuse is afforded for the use of torture in their courts of justice; for it is argued that where the value of an oath is not understood, some other means must be resorted to to extract evidence, and the readiest means to hand is doubtless torture'.[20] Often, when accusations of abuse were brought to the attention of colonial administrators, mine staff explained that 'if you want to control them', it was necessary to be 'strict' with punishments. After one riot, neither the compound manager nor the reviewing magistrate thought prison 'a feared punishment by coolies'.[21] Oriental, that is, cruel and unusual, punishment was necessary to deal with Chinese misbehaviour.

Such a view was countered by other 'experts', however, with experience from other parts of the world. Early on, one of the miners sent to China had written back warning: 'Physical violence is an abhorrence to the Chinese, and anyone offering it to them at once loses cast in their eyes, and becomes to them a despicable barbarian, and unworthy of respect. Further, an exhibition of violence might lead to the whole of the men of any mine refusing to work, and to general trouble.'[22] Mine managers and CMLIA circulars reminded its members occasionally of these points but white supervisors on the ground were more apt to treat the Chinese as they did African workers, especially when the government was initially so compliant. When corporal punishment was later criticised by the government, proponents of it increasingly emphasised the Chinese character to justify its use.

Of course, discussions were not always as clear-cut as the FLD taking the side of the Chinese and the mine management the side of the white miners, as internal debates about Chinese secret societies demonstrate. One of the few issues of accord between the mines and government were with regard to secret societies. Recent historiography has demonstrated that these organisations were often far from criminal in their activities. Many were similar to a Scottish Society or the Masons or other social networks, bound together by a shared place of origin, language, or business interest.[23] Furthermore, clans were institutions found primarily in Southern China, not Northern China, where the vast majority of the miners originated. Despite this, mine management assumed such organisations were widespread. By 1904, the image of evil, secret Chinese webs of corruption was already appearing in the popular press. While very little evidence was ever provided, the government, mines, and general public alike were convinced that secret societies were prevalent among the workers and caused large numbers of disturbances. According to the administration's 'experts', each society offered protection should a labourer be punished or run out of money, but in turn members had to obey their leaders absolutely.[24] These leaders were thought to abuse their position to extort funds, lead riots, and escape detection and represented an alternative power structure within the Chinese compounds over which the mine administration had no control and little understanding. Fundamentally, this threatened the control the mines had over their workforce. The failure to identify alleged criminals was often blamed on either clans protecting each other or rival clans blaming each other.[25] It was thought the ordinary Chinese were 'very excitable by nature, and it would be quite easy for a couple of men to excite them'.[26] Therefore, any suspected secret society activity was deemed illegal and crushed as best they could, meaning the criminal element was more likely to exist while the more innocent social networking element was destroyed.

The issues of excitable Chinese and secret societies took on a particularly sinister aspect to FLD and mine officials when they considered gambling among the Chinese. As one book described:

> Behind their dry, fantastic faces and in their hollow voices lurked a passionate regard for the great god Chance and a disregard for the sanctity of human life. Almost every Chinese miner on the Rand was a gambler. They would gamble all day long in little, blue smocked groups in the spacious Reef compounds, so impassively, so motionlessly, that it was always difficult to tell which were the winners and which the losers. But winners and losers alike had the shadow of death over them. For if a man lost consistently, and, having gambled away clothes, rations, everything for months in advance, was adjudged to be in a hopeless position, the law of the coolie was that he must die. Leniency was rarely shown.[27]

Between June and October 1906, the mine statistics attributed 21 suicides to gambling debts, although there is little way to confirm this.[28] In one case a riot was ascribed to 'old gamblers and opium smokers, who are too debilitated... and who cannot earn sufficient by means of their work to enable them to compete when gambling against coolies earning large sums at the end of the month'.[29] It was widely thought among the administration that the debts miners incurred was one way secret societies exercised control over its members and got them to commit further crimes. The Superintendent claimed 'gangs of professional gamblers have established themselves in the compounds, and are there living lives of luxurious ease at the expense of their less sophisticated countrymen'. He even described 'one poor wretch' who went to him seeking protection from such a gang, since he had become indebted to them for £200 and they were either going to kill him or make him rob for them.[30] A magistrate estimated that roughly 'ninety per cent of the murders perpetrated among the Chinese themselves were the result of this craze',[31] that is, gambling. Bagot was concerned too, warning 'that lawless individuals may seize the opportunity to attempt to obtain by violence and housebreaking the wherewithal to settle their gambling obligations'.[32] A Senior Medical Superintendent aboard one of the ships also thought 'care should be taken that skilled gamblers and desperadoes should not be shipped. These could pose as coolie recruits, and yet, by their depredations on the fellows during the voyage make enough to refund the company their expenses and pay their passage back to China where they could then live a life of luxury for [the] remainder of their lives'.[33]

Chinese police were often blamed for heading criminal 'societies' and organising gambling and other illegal activities. The original reports from both Perry and Skinner in 1903 had thought the Chinese would be unwilling to take orders from white people, and had recommended that Chinese supervisors be employed. This was based upon the assumption that the Chinese were fiercely nationalistic and would best take orders from one of their own.[34] There was the practical point too that white supervisors were more expensive. Consequently, the CMLIA incorporated Chinese police into their administrative structure. The title of 'police' did not mean that they had a legal position, merely that they were supervisors of the Chinese beneath them and were paid slightly better for this.[35] Usually, Chinese police supervised groups of 50–100 workers, although this varied between mines as it had not been specified in the regulations. Once they had arrived on the mines, the group of workers would live in barracks together, eat together, and generally work together, all with the Chinese police nominally in charge. The system was similar to that used for 'natives' and also to the system of administration of Chinese labour in the Malay state, so initially no great consideration went into adapting it to suit the specific mining situation on the Rand and, ironically, given their potential function as translators, at no point was English a requirement for the position. Almost immediately, however, the Chinese police were linked by FLD and mine staff alike to crime and secret organisations. A Chinese policeman named Li Kuei-yu, an ex-soldier of the Wei Hai regiment 'and of bad character', was repatriated because, as Jamieson described him: 'Like all Orientals, placed in a position of unfettered authority, he developed into a tyrannical bully' who had 'adopted an insolent demeanour towards every one placed over him.'[36] Solomon thought them 'an atrocious lot of scoundrels and the sooner they are got rid of the better'.[37]

While there is almost no supporting evidence for these criticisms of Chinese police and secret societies, there was a petition of 230 Chinese men, a few articles in Chinese newspapers and various sworn oral testimonies from investigations which included accusations of Chinese police beating and extorting labourers, selling opium, preventing complaints being put in petition boxes, and forming a secret society 'for trading in opium and gin, compelling coolies to join a theatrical company and starting a gambling hell wherein coolies are forced after every pay day to gamble away their wages, with the result that suicides by opium poisoning are frequent'.[38] It seems likely therefore that there was some abuse, but despite their apparent unpopularity nothing was done to replace or reform the policing system.[39]

Instead, administrators decided to ban gambling completely, describing it as 'subversive to the maintenance of order and discipline'.[40] Before the end of the month, the government gave the Lieutenant-Governor powers to ban gambling in the Transvaal Ordinance 17 (1904) update of 1906.[41] This approach to 'fixing' the gambling issue reflected the inherent problem with viewing the Chinese as 'working machines'. The Chinese 'have practically no distraction after their day's work is over, save gambling'. The mines were boring, without even 'wholesome' religious activity available. Even 'Taotai Lew, the Chinese Consul General, who is himself a Christian man' expressed 'his astonishment that so little religious work is done among the Chinese.'[42] It was only when a small number of Catholic Chinese specifically petitioned the FLD to be allowed services that the government relented, much to the annoyance of mines, who resented having evangelists on their premises.[43] Instead, the mines increasingly applied measures meant to control the social lives of African miners, as well as enacting new legislation. 'Natives' were legally banned from drinking alcohol in 1902 and this was applied to the Chinese without warning, until a labourer was arrested and successfully prosecuted.[44] Initially, to aid recruitment, the mines allowed the sale of opium on the gold mines for medical purposes,[45] but it too became associated with the 'panic' and so was banned. The archives make it difficult to determine how widespread or dangerous the activity was. On mines and voyages, men from Chinwangtao and Chifu were particularly prone to being listed as dying from 'opium poisoning'[46]; although contaminated supplies might have existed, prejudices about opiate users and Chinese hygiene almost certainly played a factor in its mention as a cause of death. It remained one of the leading causes of death on the mines among the Chinese, with 6.83 dying on average a month between December 1904 and November 1906. If a user died or became ill, no other health factors were considered. Opium was frequently cited as the cause of death in cases where opium could in no way be responsible. During one ship voyage to Durban, a Chinese man allegedly died of 'influenza and jaundice' and two more of diarrhoea 'which turned into dysentery'. The medical official blamed their 'addict[ion]' to opium as the causes of death in all three cases, despite the fact that opium in large quantities causes constipation, not diarrhoea, and certainly would not cause influenza or jaundice. It is even possible the men with diarrhoea were self-medicating, and the actual illness was due to unclean food or water supplies on board ship.[47] This was nowhere mentioned in the report; instead the focus was on controlling the 'typical' Chinese behaviour on board ships.[48] The assumption that opium smoking was a normal Chinese pastime was common, as was a belief in their lack of fear of death. The only reported incident of a clear link between the existing Chinese population and the labourers was when a Cantonese man named Long Jak, a storekeeper at Simmer and Jack, was charged with having 7½ lbs. of opium for sale to the indentured labourers.[49] In response, TEAs in China were encouraged to ban opium and confiscate pipes when possible,[50] while the Transvaal Law Department drew up legislation to ban sale of the drug to any Chinese.[51]

Because of continuous public scrutiny and their own desire to make Chinese labourers as efficient as possible, the mines and the FLD increasingly implemented restrictive legislation which made socialising almost impossible for the Chinese and were probably counterproductive, exacerbating desertion and other problems. Confining increasing numbers to the mines was likely to breed boredom and unrest, nor was it easy to enforce. While the mines at Kimberley were closed, making it easy to keep track of staff and to largely prevent the theft of diamonds, the gold mines had never operated along these lines; generally the sites were much bigger and the boundaries somewhat haphazard. Compounds were often surrounded by a fence, with access through one or two gates, kept open during the day and sometimes at night as well. The compounds themselves varied from 300 to 1200 acres in area, covered with numerous buildings, slag heaps, and the mines themselves. Such a large and busy area was almost impossible to police effectively, but increasingly, attention focused on how to do just that, as well as how to segregate the Chinese from Africans.

While some mine conditions improved because of the Chinese arrival – the availability of hot water and more spacious and better ventilated accommodation,[52] for example – other aspects became worse. Throughout 1905, there were increased Africans recruited but not enough jobs for them to fill, so wages dropped. This might have helped feed into the violence outlined in Chapter 5.[53] Because administrators inevitably interpreted such conflicts as racial, their solution was segregation. After one disturbance, the mine involved not only moved the native location, they also discharged all the Africans employed by the company on the mine, apart from the 'boys' working on a Mill and Cyanide Works, who were then kept away from the Chinese.[54] On mines where the elimination of African labour was impossible, it became usual to have all Chinese and African miners work separate shifts in order to still ensure as much segregation as possible.[55] The results of this discourse would be to firmly enmesh a system of racial segregation within gold mining, later applied to all aspects of South African life. This was not simply a matter of control, of divide and rule, but reflected an easy way to address conflict without having to deal with the more complicated issues raised.

The blame game

To the administrators, both from the mines and the government, trying to define the perceived 'problem' behaviour within the parameters of the 'Chinese character' and passing laws to regulate such behaviour and to control the physical bodies of the Chinese men was their attempt to exert control over the situation, over both the public hysteria and the Chinese themselves. However, as time went on, and public scrutiny more and more critical, the attention of the FLD and the mines shifted from debating the 'Chinese character' to just generally blaming each other for any perceived failures to achieve 'control'. To solve the problem of corporal punishment, for instance, Selborne and Jamieson focused administrative attention increasingly on white mine staff, blamed for outright cruelty, incompetence or simply failing to invest in the compound infrastructure as needed. While not denouncing all mines, Selborne was clearly shocked by the extent of the mines' 'amazing carelessness combined with the most shortsighted parsimony in respect of management...Chinese outrages...are, in my opinion, mainly if not wholly, due to this mismanagement'.[56] Excessive drinking and other behaviour which was 'discreditable' to whites was a frequent criticism.[57] Selborne wrote a memorandum which dwelt at length on this and recommended 'the provision of a sufficient supply of good houses for married miners, immediately adjacent to the mine on which they are working'. This would encourage men to bring their families with them, and women could thus exert their civilising influence; a more stable, less violent workforce, it was assumed, would then materialise.[58]

On the other hand, many white miners had never wanted the Chinese and found working with them very difficult, since many of the Chinese did not, in their views, follow basic safety measures when handling dynamite, which endangered them all. White miners were also used to treating Africans violently, so generally did not understand objections from outside the mining communities. Instead, they thought government interference, especially the banning of corporal punishment, endangered them. One mining engineer told the *Illustrated London News* 'If the Chinamen broke out and threatened the whites or native boys, we were ordered to use no violence, even though our lives were threatened'. He also described the Chinese in suitably racialised terms, as people who loved the 'deviltry' of violence against whites for its own sake and who had 'no moral sense at all as we Westerners see things'. His point was clear: the only way to deal with such 'deviltry' was through violence.[59]

The mines and government officials met several times to discuss the matter. In December 1905, a deputation of Bagot, Langerman, Brakhan, Chaplin, and Webber (representing all six major companies in the Chamber of Mines) had met with Solomon and Jamieson. Bagot thought 'the establishment of *white* watchmen on the mines...will result in a large reduction of the absentees if a proper system is thoroughly carried out'.[60] However, the mines were reluctant to make this a requirement, given the expense involved. Instead, the group decided 'that the Superintendent of the FLD should be given discretionary powers in the case of poorly managed mines where the management and control of the Chinese labourers was unsatisfactory or deficient, to insist on the engagement of a white man speaking the dialect of the coolies'. In return, Jamieson promised not to interfere 'in the case of mines having no white controller where everything was proceeding satisfactorily'.[61]

Mines were willing to admit to the problem but were increasingly unwilling to implement costly remedies. From his position as the most senior Wernher, Beit & Co. executive in Johannesburg, Lionel Phillips agreed privately that lower-class white men were a problem. He agreed that more married men would 'undoubtedly' help, and he wished to increase the society and entertainment available to whites, thus 'raising the tone of the [white] working man'. While some moves were made in this direction (in 1902 only 20 per cent of married miners had their families with them; by 1912, it had risen to 42 per cent),[62] the change was slow. The mines had laid out a great deal of capital on the original terms and conditions of the Chinese scheme, in addition to the reconstruction, expansion, and improvements made after the war. Realistically, he wrote, they could ill afford additional housing expenses. Investors were unwilling to back such schemes; 'the moment is not propitious to consider any large and avoidable capital outlay'.[63] His comments were more than an excuse. The Liberal electoral victory had brought with it a dramatic slump in gold mining share prices, and many of the mines were even more financially precarious than in 1902. Even the bigger ones like Corner House and ERPM were struggling to recoup their expenditure on Chinese importation or match the returns expected by investors.[64] The situation, coupled with the continued political uncertainty, meant that they were reluctant to use the little capital they had to implement expensive policies, however it might have improved mining conditions. Chinese labour, it was turning out, could undermine the very market confidence and security which they were eager to establish.

Overall, despite some agreement about white working class miners, rather than the conciliatory attitudes found at the beginning of the experiment, the mines now felt themselves to be unfairly persecuted and blamed for matters beyond their control, while being taxed severely. Relations had clearly soured when Farrar wrote angrily to Selborne, criticising 'the unfair interference of His Majesty's Government with the rights of private contract under an ordinance which has been sanctioned by His Majesty'. He admitted 'there has been slackness' but blamed the government's ban on corporal punishment, not the mines. It was 'very difficult indeed to exercise as much control as we should wish in view of some of the clauses of the present Ordinance'.[65] In other words, rather than a partnership, increasingly mines blamed government's interference in their affairs for problems and the government blamed the sorts of white men employed on the mines and a general meanness with money. Both saw the remedy in more regulations, just different types of regulation. In 1905 and 1906, Farrar used his position in the Transvaal Legislature to push for amendments which would allow collective fines and for all supervisors or police to be legally responsible for the men under them unless they reported any offences.[66] The mines also wanted it to become criminal for 'labourers inciting other labourers to infringe the Ordinance'.[67] The government, however, refused these suggestions, further infuriating the mining industry.

Desertion was also a frequent point of friction, as the public blamed this for most of the attacks and sexual 'vice'. There was even an industry of making maps for the Chinese to be able to 'walk' back to China.[68] Again, attention shifted during the panic from why the Chinese deserted to how to control them. In February 1906, Jamieson felt compelled to write to the President of the Chamber of Mines, about 'whether or not it would be advisable for the mining companies to place police guards of their own around the compounds at night', as there was 'an increasing number of coolies wandering about in the country' and 'the feeling amongst the farmers against the Chinese is growing'. He made clear the government did not want to give 'each compound the appearance of an armed camp', so was reluctant to use the SAC or local constabulary for the job.[69] Fencing was also considered, along with tighter controls on issuing passes. More frequent roll calls would also enable the mines to know if a Chinese worker had absconded more quickly, making their capture easier. This would cost money, however, and neither the mines nor government bodies could agree on exactly which measures were best, or who would pay.

Among the most senior levels, there were some attempts to reach a rapprochement, but these had little wider impact. Selborne wrote to Milner, frankly explaining his views:

> [T]he *whole* blame for this business is the folly and incapacity of some of the mine managements... The Government have done everything in their power in the matter... nothing has been able to rouse certain of the mine managements to a sense of their responsibility... The longer I am here the more impressed I am with the amazing amount of inefficiency and incapacity there is among the mine magnates. It all arises from this cursed absenteeism of the really responsible people.[70]

After all, people like Chaplin, Farrar, and Phillips travelled frequently back and forth between London and Johannesburg, and Farrar was head of a political party in addition to a mine magnate. Whatever care they had taken in laying out the regulations for the scheme, whatever concessions had been made, they could not supervise the mines personally, nor desired to do so.

A further problem was that the administration on the mines had never been uniform. The governments involved had worked chiefly with the Chamber when forming the regulations, but the Chamber oversaw directly only recruitment and transportation. Once the Chinese had arrived, the administration was subject to the wide variety of organizational structures within different mining companies and among mine managers. Even within one mine, individual compound managers could have varying administrative strategies (see Table 4.2: CMLIA Membership and Table 4.4: List of Mines utilising Chinese labour). Some mine managers were better prepared, some translators better than others. A small number of white miners, particularly those unionized, were keen to see the experiment fail and so leaked unfavourable stories to the press or continued to organise anti-Chinese campaigns. Colonial government officials might increasingly wish to standardise mine practices on the Rand, but there was no effective way to force the matter. While still arguing that the mines and capitalists were developing a closer relationship during this period, even Yudelman has admitted that 'individual mine managers' had 'a large degree of relative autonomy *vis-à-vis* the mining houses'.[71] The mines had to be flexible in their administrative structures, to deal with the varying mine conditions and sizes; this made it equally difficult for the colonial government to impose regulations on the mining companies. None of this stopped the government from trying and the mines from resisting. Consequently, trying to fix the administrative structure for all mines was bound to be difficult. The political situation made this especially so.

The agreed banning of opium, gambling, and the like, or the unanimous but ineffectual agreements that something needed to be done about white supervision actually perpetuated the very problems they sought to stop, leading inevitably to more problems and continued friction. The most frequent complaint was about the degree of power the FLD held over mine management. The mines constantly tried to restrict the FLD supervisory powers, insisting that it was 'subversive of mine authority, or discipline'.[72] Someone from Corner House requested that 'any enquiries instituted should be held in the presence of some responsible official of the mining company concerned, and the inspectors should at all times inform the manager at the mine of their arrival on the property.'[73] This was a standard wish of the mining companies;

they wished to cooperate but were constantly worried about anything which might undermine their authority in the eyes of the Chinese. And as problems continued to receive so much public attention, the mines increasingly complained about the FLD which, in their view, let disturbances 'drag on for days whilst negotiations are being carried on with the Coolies'.[74] Corner House directors complained that they were prevented from 'getting as much work out of them as we could have got' because they were 'crippled by the conditions of the contract, the regulations imposed by the Government, and the interference of the officials'. They argued that as housing and food were so good and free to the Chinese, the workers did not have sufficient motivation to work, while dismissing the moral panic after Joubert's murder as 'hysterical'. As was often the case, the discussion turned to money. Their biggest concern was to ensure that the murder not be 'made the excuse for further burdens imposed on the industry'. They were concerned about popular suggestions to make the mines pay for prison accommodation, that the food in prison was too good and the Medical Officer of Health for the Transvaal was a 'faddist' and too slack. And despite the many ways the government tried to assist the mines, they opined that the 'authorities in Pretoria' were not 'more sympathetic' to the mining industry.[75] The CMLIA end-of-year report in 1905 made similar comments, although the continued public outcry was taken a bit more seriously in this public document than in their private correspondence. The destruction of a store near Germiston with dynamite was 'not yet proved' to be 'the work of Chinese labourers', but they assumed it was, or at least that they would be held responsible by the public. As such, they were concerned about how incidents 'prove a powerful weapon in the hands of those who advocate the repatriation of the Coolies'. They recommended that they be allowed to search local shops, especially those of 'lower class store-keepers' who were suspected of buying stolen supplies, including dynamite, from the Chinese on the mines.[76] They did not consider how this would be received among the local population, however, and it was not allowed.

Sometimes there was an entire breakdown in relations. When Jamieson and Solomon jointly asked the mines for information as to how many Chinese were employed on mines in non-mining capacity as cooks, gardeners, launderers, and so on, the CMLIA responded: 'under existing circumstances it is considered inadvisable that any information should be given by this Agency unless it is perfectly clear for whom and for what purpose such information is required'. As Solomon and Jamieson were the two most senior government officials responsible for the scheme, clearly political and personal conflicts had frayed relations considerably. Indeed, the information was only handed over when the head of the CMLIA was shown the cable from the Secretary of State for the Colonies requesting the information.[77]

One of the frequent matters of conflict involved the ever-rising costs of the scheme. Some aspects of recruiting and housing the Chinese had been specified in the initial regulations, but problems arose with the increased policing expenses as a result of the 'outrages'. Intense resentment was felt on both sides over whether compensation should be paid to victims of alleged Chinese assaults. Initially, the government expected that 'any specific expenditure' incurred because of the labourers would 'be paid out of revenue derived from the passport fees'. But almost immediately, the Acting Commissioner of Police had to increase police numbers to deal 'with the disturbances on the mines caused by Chinese labourers', costing approximately £4000 a year. Furthermore, the FLD decided it was 'not advisable' to put Chinese and African prisoners in the same jails, so wanted new ones build just for the Chinese at mine expense.[78] The original passport fees were simply not big enough to meet such expenses. When asked for more money to help with policing in early 1905, the Chamber insisted that it 'comes properly within the general provisions for maintenance of law and order and should not be derived from special revenue whether passport fees or any other source'. Instead, the Chamber agreed passport fees were too high and asked the government to lower them.[79] Such disagreements were frequently as much about money as they were about the balance of power between the government and mines. In 1906, when the FLD insisted that their courts could not be held in compounds but that the mines had to pay for new court houses to be erected nearby; the FLD was trying to ensure justice appeared uninfluenced by the mines. The mines, however, complained about the extra bureaucracy but mostly about the extra cost involved.[80] As late as April 1905, Selborne was hopeful that passport fees would eventually cover the extra expenditure,[81] but neither the government nor mines fully recovered their costs.

The results of this can be seen in the debates at one CMLIA Board meeting discussing how to reduce absentees and minimise 'outrages'. While there were agreed measures, they were all rather tame. No permits were to be issued to any Chinese after sunset on Saturday evenings, nor were they allowed to leave the compound again until 9 am on Sunday mornings. Police would be asked to arrest any outside mine premises during those hours. No permits for more than 24 hours were to be issued and their destination written on the permit, only allowing them to travel between the two points. They also wanted punishments for desertion or illegal absence to be more 'severe'. They were unable to agree upon the cause of absences or desertions, 'employment of white mine police in lieu of Chinese police', or building of barbed wire fences, given the extra costs involved, or whether to discontinue all overnight permits.[82] There was also tension over the number of staff. While the mines often requested cutting staff to save money, the FLD often complained about not having enough staff or the staff being of insufficient calibre.[83] The FLD wanted to improve the quality on board ships, to hire more ex-soldiers and the like to act as interpreters or Chinese police, and generally to secure a better class of men. The CMLIA dismissed these ideas and instead wanted fewer medical staff.[84] Eventually Lyttelton was asked to mediate and refused to sanction a reduction in medical staff but did not require more staff being hired.[85]

Even when the mines and government seemed on the surface to work in relative harmony, money often revealed the acrimony underneath. After a relatively amicable 1906 Committee met to discuss possible ways to better control the Chinese, arguments began over who should pay the £924 for publishing the report.[86] The Executive Committee of the Chamber of Mines insisted that, since 'the inquiry was undertaken in the urgent interests of the public', the Colonial Treasury should pay.[87] Solomon, however, stressed that the meeting was only necessary because 'it is the duty of the mines so to control their labourers on their mine premises as to prevent them becoming a danger to the community', and they were failing to do this.[88] The Chamber eventually agreed, but still insisted that 'the whole inquiry was undertaken in the public interest' so 'the costs should be paid out of public funds'.[89] In turn, the government eventually paid for the copies sent to them 'at the price at which they are sold to the public'.[90] Such issues should have been minor but some of the most senior government and mining officials spent a month bickering about this matter, reflecting the animosity behind the scenes, whatever public facade they might have given. It also meant that solutions became more and more difficult to negotiate.

Conclusion

While adaptations to the scheme were designed to eliminate public criticism of the scheme, the political situation increasingly made the scheme moribund. The failure of the mines and local government to be able to work together, their insistence on criminalising most Chinese social activities and the continued reports of violence and immorality made the failure of the scheme inevitable. Through a series of debates over the 'Chinese character', most mining employees and government officials came to understand them prone to Eastern despotism and violence, filth and disease, and hence in need of constant supervision and regulation, in order to ensure they worked efficiently and did not escape to commit 'outrages'. Changes to the administrative structure were meant to reassert mine and government control over the public hysteria and over Chinese workers alike, but failed to do either. Given the administrators' preoccupations, only a significant financial investment would have enabled them to 'control' the Chinese. As the British government often found when it considered ruling by force, the finances simply were not there. Rather than aiding the building of a stable white settler colony, both blamed the other for obstructing this goal.

As the previous chapter explained, the motives for Chinese actions in such incidents were difficult for the administrators to fathom and remain unclear in the records. Did they desert because of the brutality they suffered or because of criminal elements among the Chinese imported? How much were their daily lives defined by violence, or other forms of exploitation, or were there other factors at play? Were they just bored? The heavily politicised nature of public discourses, and acrimonious discussion in private make the archives unreliable, allowing only speculation. What is present in the archives reflects how many images were created of the Chinese 'character', while at the same time colonial and mining administrators' records 'silenced' the Chinese, interpreting Chinese actions and motives for their own purposes and suffused with their own race and class prejudices.[91]

The frequent archival portrayal of the Chinese as pawns of capitalists, of party politics, of white (often male) insecurity, of Afrikaners trying to undermine the British, or even of a few elite Chinese police, deprived the Chinese of agency in both senses of the term: they almost always lacked any individuality in the records, while their agency to act was depicted as largely confined to violent protest. Which in turn was not about individual power to act but was usually blamed on irrational racial characteristics or corrupt Chinese (or occasionally working class white corruption). This denial of Chinese agency also meant that administrators thought that they could fix any problem by controlling Chinese bodies; other options were not considered. Nothing in the prolonged negotiations over the terms of indenture was prepared to deal with the moral panic which erupted in the Transvaal, nor the global coverage it received. This situation put enormous pressure on the mines and government overseers of the scheme to adapt the administration to better demonstrate that they were in control. Whether the Chinese were ever really out of control ceased to be the point in such an increasingly politicised and acrimonious environment.

This was not just a case of one side having power and the other not. The mines and government sought to gain control of the situation and failed. They failed to reassure the public and they failed to minutely control the Chinese bodies, or understand the Chinese 'character' as they wished to do. They fought among themselves. The Chinese retained some ability to go on strike for better conditions, or to act in other ways counter to colonial control, right up until the last were repatriated in 1910. Yet the mines did gain certain powers, while the level of government oversight of mining affairs actually increased. The establishment of this administrative mindset therefore had significant ramifications, not just in politics, but also in creating a legislative mindset which fostered segregation and state control, two of the most important building blocks of apartheid.

7
Political Repercussions: Self-Government Revisited

As had always been clear, any decision about Chinese labour in the Transvaal would have wider repercussions. Because of the moral panic and failure of the mines and government to gain 'control' of the situation, there was a repeat, in many ways, of the debates outlined in Chapter 3, once more directed not just at a local audience but at an imperial one. The chief focus of discussions centred once more on the role of the imperial government in determining the policies of individual colonies. There was again the pull of grand ideologies like whiteness, democracy, imperial federation, and anti-capitalism played out alongside unique, increasingly *national*, concerns in these colonies. Once again, these issues were played out in the press, through public meetings, transnational organisations, and elections. There were petitions but far less of them. But unlike the debates before the importation of Chinese labour, imperial federation was hardly mentioned, the gloss of the partnership of the South African War having faded. Instead, white male self-government was emphasised by all sides of ensuing debates over Chinese labour, with non-whites increasingly marginalised.

The 1906 British election

Global interest had faded somewhat when the scheme had first begun but the moral panic in the Transvaal and the British accusations of Chinese slavery ensured that the scheme appeared in newspapers weekly throughout 1905–06. This was not just because of the panic but because by mid-1905 it was clear that a British election would be held soon, and that the anti-Chinese Liberals would win, with Chinese labour featuring heavily in their election campaign. In the run-up to the election, comparisons were made between Eastern European and Chinese migrants:

both allegedly stole British jobs, lowered wages, and imported disease and crime.[1] Such a comparison was particularly effective because false rumours circulated in Britain that the government was planning to import Chinese into Britain too, to undermine growing labour movements. The widespread interest in, and fear of, Chinese labour was a significant shift in attitude from even a few years before. While the 'white' colonies had debated the issue of Asian migration for decades, Britons had felt themselves largely untouched by the issue. Only a few years before, the CO had said of the Natal language test: 'The shoe doesn't pinch us; for in the first place each Asiatic in Natal must be multiplied by eight hundred to produce a proportionate effect on the population at home; and secondly this country being already fully populated, a relatively large influx on a foreign element could only be brought about by a corresponding displacement of the native element.'[2] For the first time, Britons felt themselves to be threatened too: 'while a kaffir could never take the place of a skilled miner, the Chinaman will be able to do so. The opinion is freely expressed that in six months there will be plenty of Chinamen well able to replace the skilled Cornish miner'.[3]

The Liberals also benefited from the close cooperation of fringe labour and socialist parties, united against Chinese labour. In the run-up to the election, 67 per cent of Labour Representation Committee (LRC) addresses and 36 per cent of Socialist speeches mentioned Chinese labour, although both generally avoided international issues. The LRC 'associated both the war and introduction into South Africa of Chinese labourers with the enrichment of a privileged minority'.[4] The ASE and the British Trade Union Congress also wrote and published pamphlets and articles, gave speeches, and signed petitions against the proposal, with campaigners travelling frequently between Britain and South Africa. Indeed, so effective was the campaign, both organisations passed resolutions banning any Chinese from joining their ranks, despite the complete lack of any Chinese trying to join.[5] Perhaps most importantly, after ten years of power, people were sick of a party now linked so strongly with 'Chinese slavery', with men campaigning in the streets dressed as 'pigtailed and manacled Chinamen'.[6] The Liberal Landslide of 1906 saw the Unionists lose 60 per cent of their seats, down to 157 MPs, while the Liberals won 377, giving them a clear majority.

Repatriation

It was unprecedented for a colonial matter to gain such prominence in a British election and it led to serious imperial complications. Britons expected the new Liberal government to act quickly to resolve the matter. This was problematic, however. Should the imperial government settle the Chinese issue right away by banning it, or was it better to establish self-government and let them decide for themselves, which would take time but be more democratic? For some, the reports of Chinese 'outrages' and Chinese 'slavery' fuelled the argument that immediate imperial action was needed,[7] while for others, that was how the whole problem started, and the imperial government needed to stop interfering and let democracy decide.[8] And of course others still did not want the scheme cancelled at all. The debate over whether the scheme should be cancelled followed by now familiar divisions. Those wanting to continue Chinese labour were largely based around the Rand mines and organised by senior mining figures. Those against were pretty much everyone else, although again, the mines were clearly the best funded and organised in their efforts. Once more there was a debate over who should have a say: only the Transvaal? All of South Africa? The entire empire? It is perhaps unsurprising that one of the first acts of the new Liberal government was to begin the lengthy process of negotiating the terms of responsible government as 'the only way out of the impasse' over Chinese labour.[9]

Sorting out the terms of responsible government would take time, however, so the British government unsuccessfully asked the mines to voluntarily stop immigration for six months.[10] They decided again cancelling outright, largely for electoral rather than political reasons, as they would have had to reimburse the Chinese and the mines for forcing the contracts to end before their completion dates. The immediate fall in gold mine share prices after their electoral victory further warned the Liberals that the situation was a sensitive one. Particularly in France, the Liberal meddling was blamed for poor investment returns. Still, the use made of the issue in the election ensured action had to be taken or deep embarrassment suffered. In February 1906, a Unionist member moved an Amendment to the Address, regretting that 'Y[our] M[ajesty]'s Ministers should have brought the reputation of the country into contempt by describing Chinese indentured labour as slavery and yet are contemplating no effective measures to bring it to an end.' While the Amendment was defeated by 416 votes to 91,[11] demonstrating their parliamentary dominance, it was also embarrassing, which made them eager to resolve the matter quickly.

Consequently, the British government suspended the further granting of importation licenses, 'pending decision as to grant of Responsible Government' and put in place a voluntary repatriation scheme.[12] The imperial government pledged to cover one-third of the cost if labourers volunteered to return home and found the additional expense for repatriation themselves.[13]

Backlash over 'interference'

While many wanted the repatriation of the Chinese, the imperial government's decision led to fierce debates. Even before their enforced cessation, there was campaigning from the mines, who unsurprisingly argued that the imperial government should not interfere and that Chinese labour should continue. Once again pamphlets and books were published, speeches given, and letters written.[14] Two mines actually started operating daily tours open to members of the public for a small fee, to demonstrate to the public that the conditions of labour were good, the situation under control, and to satisfy the curiosity of visitors from all over Europe, the United States, southern Africa, and many parts of Australasia, an amazing tourist phenomenon most comparable to the World Exhibitions at the time. Some of the larger mining companies also began issuing their own staged photographs and postcards to show Chinese life. At one point in 1906, a mining company even considered paying for a short film to be made so that people in Europe could see how well treated the Chinese were; if a film was made, unfortunately there is no record of it,[15] nor is it clear which were staged images paid for by the mines and which were spontaneous snapshots of life. The Chinese 'spectacle' had a distinctly political edge.

Efforts really took off after the Liberal victory was secured, with the Chamber holding a special meeting of the Executive Committee to condemn government interference within weeks.[16] After the license freeze and repatriation were announced, men from several mines passed resolutions requesting the High Commissioner allow the continuation of Chinese workers, or there would be high white unemployment.[17] Similar public meetings were held in Benoni, Boksburg, and Nigel (all with big Chinese contingents and with jobs dependent on the mines). Roodepoort and District Chamber of Commerce, Johannesburg Town Council, the Transvaal Association of Mine Managers, Johannesburg Chambers of Commerce and Trade (Incorporated), and the Pretoria Chamber of Commerce all did the same, as did the South African Operative Masons' Society branch in Pretoria and the Rand Pioneers.[18] A conference was organised from various Rand commercial organisations, including the Chambers of Commerce of Johannesburg, Boksburg, Germiston, Krugersdorp, Randfontein and Roodepoort, the

Johannesburg Chamber of Trade, Transvaal Mines Traders' Association, Central Rand Storekeepers' Association, East Rand Storekeepers' Association, and the West Rand Storekeepers' Association. The meeting unanimously resolved that 'ample opportunity should be afforded the commercial community of considering the terms of any new notice which the Government proposes to issue'. The Johannesburg Chamber of Trade in particular objected to the terms of repatriation since it lowered 'the dignity of white men in [the] eyes of yellow and black races... from [a] moral standpoint interference between [an] employer and employee in the matter of [a]signed contract cannot be justified'.[19] Indeed, it was unprecedented for the British government to interfere in specific industrial issues in this way, but such overt interference had been part of the scheme from the beginning.

Most of these complaints focused on the rights of Transvaalers to decide the matter for themselves and on the economic consequences. Many white miners were clearly scared of losing their jobs if Chinese labour ceased. The mines certainly fostered this view and mine managers showed leniency in allowing time off to attend meetings in support of Chinese labour, organising several such meetings to occur on the mines themselves. One American miner wrote home expressing his frustration over the ensuing uncertainty: 'People are actually looking now for the worst, which is the closing of the mines and the financial ruin of South Africa... If this is the way England treats her colonies, all I have to say is, I am sorry I did not live in 1776, but, perhaps, I will have a chance yet.'[20] While such feelings of rebellion were uncommon, tensions were extremely high.

Most protests combined economic concerns with complaints over interference from London. On 15 January, the Krugersdorp and District Chamber of Commerce passed a resolution protesting 'against the misrepresentation and calumnies urged by party politicians in Great Britain with reference to the Chinese labour', which they considered 'essential' to the local economy. Similar resolutions were passed by the Germiston Chamber of Commerce; the Johannesburg Chamber of Commerce wished the matter to be decided only by responsible government.[21] Farrar told Selborne he thought the idea of repatriation 'a question of gross... interference' by the government[22] and in the same month, the Chamber and Germiston Town Council again cabled resolutions against repatriation as detrimental to the economic well-being of the area and harmful to white employment which, they argued, was dependent on having Chinese unskilled labour.[23]

While such views were well organised, the majority of the Transvaal (and elsewhere) were against Chinese labour remaining in the colony. This time, Afrikaners, largely (although not exclusively) through the newly formed political party, Het Volk, wanted to leave no doubt of their views and petitioned the colonial and imperial governments frequently, usually referencing the Chinese desertions and 'outrages'. Similarly, a Cape-organised South African Liberal Association held meetings in southern Africa and Britain in 1905 to ensure both that Chinese labour remained an election issue in Britain and that Transvaalers remembered it was potentially a 'national' issue for 'South Africa'.[24]

Once again, there was widespread transcolonial attention, but it was different. There was a regular stream of coverage in Australia and New Zealand in particular, but newspapers increasingly copied cables from London with little commentary. Part of this was a general lack of surprise expressed over the 'outrages'; it was what they expected when whites and Chinese mixed (although they still carried the salacious details). Another factor was the fact that colonial newspapers had not retained individual correspondents in the Rand since the South African War. By 1905, when the panic ensued, most coverage came indirectly from London correspondents or from Reuters or Press Association cables. One of the chief results of this method of reporting was that the newspapers (and politicians) were less clearly opinionated on the matter. They would report comments from both those for or against Chinese labour, often without any commentary expressing opinion. Colonial politicians as well were less vocal than before.

The few times local petitions were set, their reception was damning: when Adelaide, Australia passed a town resolution asking the imperial government to end the scheme, the *Scotsman* called their actions 'ludicrous' and thought their 'impertinence is downright appalling, and is only equalled by the sense of self-importance, self-sufficiency, and ignorance which leads them to think that the Home Government will take the slightest notice of their protest'.[25] *The Times* too told Australians that they should remember not to interfere.[26]

When discussing whether Australia should comment on Irish Home Rule, *The Sydney Morning Herald* felt that 'the tendency to intervene in a partisan way for one political end or another is one that ought to be steadily discouraged', although they had been justified in their comments over Chinese labour in 1903 and 1904, there was no sense that they should comment again on the Transvaal situation.[27] Richard Jebb in Australia argued that the Liberal Party's stance to end the Chinese labour scheme was no better than the actions of the Unionist government in 1904. Both imperial governments had dictated colonial affairs rather than granting self-government. When he was interviewed by the Wellington *Post* while visiting, he made clear he thought that New Zealand had also interfered in a self-governing colony when they had organised petitions in 1904. In his view, New Zealand was denying self-government to the Transvaal; he would never 'dictate a policy to a colonial Government'.[28] He warned that such actions had consequences for all the white colonies and their debates with Britain about imperial federation.[29]

Such a view was increasingly widespread. The imperial government had so 'hopelessly antagonized the British section of the community' in the Transvaal, it had 'imbued the whole population, including the Boers', with such 'contempt' for its 'party methods' and 'distrust of its entire Colonial policy', that 'no regard for its wishes or its policy need stand in our way in our efforts to save our own position', complained one of the most jingoistic mining magnates.[30] Olive Schreiner noted: 'It seems that the Almighty has ordained that *no* Englishman, whether he be jingo or Pro-Boer, should ever understand any things about South Africa.'[31] A labour-supporting newspaper, while extremely critical of the mining magnates, stated: 'The Home Government cannot rebuild South African society. Such a work cannot be carried out from Downing Street.' They did not want the imperial government to cancel the scheme, they wanted to decide for themselves. The only solution was self-government to be followed soon after by South African federation.[32] So common was this view that on 5 March 1906, the Mayor of Johannesburg oversaw a politically neutral meeting of 38 organisations (some scientific, religious, racial, and professional bodies) that demanded an immediate grant of responsible government.[33]

One New Zealander living on the Rand wrote to his brother that Seddon and other politicians abroad should 'leave these matters to be settled by the people of the Transvaal' because it was 'a subject that they do not understand'. They should be left along 'to manage our own affairs'.[34] And indeed Seddon did not organise any petitions to the imperial government this time, although he mentioned the Chinese scheme when defending his invitation to America's 'white fleet' and during his many public speeches against Britain's political alliance with Japan. If New Zealand did not stand up to Britain on such matters, Britain might treat them as they had treated the Transvaal and ignore public antipathy to Asians.[35]

There were occasional meetings, usually of labour figures, which passed resolutions supporting the anti-Chinese policies of the British Liberals 'for the sake of the honor and good name of the Empire'.[36] A New Zealand newspaper deplored the fact that 'Chinomaniacs' did not want anyone but the Transvaal to decide the matter as 'the question is one for the whole Empire to settle'[37] but such views grew rare and no mention was made of imperial federation. Instead, they were more likely to stress that they had 'earned the right to a voice in the settlement by the blood of our sons shed in a cause which is to-day regarded with grave misgivings through the base surrender to capitalist greed which rendered Chinese immigration possible'.[38]

While such debates continues, the imperial government required mines to post notices outlining the freeze on importation and the voluntary repatriation scheme in May 1906, and the FLD travelled to each to ensure the signs were up and to explain the terms to the Chinese. After all of the controversy, however, initially only 60 Chinese applied for repatriation in May 1906. Even by July, the number of applications was only 128.[39] This was a small fraction of the more than 47,000 Chinese residents on the mines at the time. There were scattered attempts to blame the mines for the low numbers, but the FLD took great care to ensure all Chinese labourers were familiar with their rights. When interviewing Chinese, the FLD inspectors found it difficult to find labourers who wished to be repatriated but could not find the funds. The jobs were better paid than most unskilled labour in China. Between May 1906 and December 1909, only 830 Chinese were repatriated under these terms.[40]

Transvaal party formations

Despite the minimal impact of the Liberal policies, the tensions between imperial and colonial decision-making had a significant impact on the development of political parties and the drawing up of a constitution in the Transvaal. The Transvaal Responsible Government Association later renamed the National Democratic Federation was the first political party in the Transvaal after the war, formed in late 1904 by a group of friends and acquaintances against Chinese labour, and officially led by E. P. Solomon, brother to Richard Solomon.[41] The group was incredibly diverse, made up of colonial officials (and ex-ZAR officials), diamond mining magnates, white labour supporters and small-scale businessmen.[42] Richard Solomon even advised them on how to draft responses to the imperial government regarding constitutional debates, giving them a prominent voice in the global coverage of the scheme, often portrayed as the official opposition to Milner's appointed legislature.[43] Despite this prominence, the party was only loosely organised. Their few official policies were often issued by E. P. Solomon without obvious consultation with other members.[44] While they disagreed on many issues, they were united in their belief that the imperial government was the great obstacle to banning Asiatic immigration. This was part of their reason for the name, the Responsibles, and later for becoming the Nationalists. It was meant to indicate that the Transvaal should be allowed responsibility for its own government policies and be granted self-government by Britain. This is why, despite their opposition to Chinese labour, when the Liberals announced their plan to repeal the Ordinance allowing Chinese importation in March 1904, E. Solomon complained of the 'interference in our affairs by English politicians... these busybodies know little of the circumstances of this country and only use us to further their own ends'.[45] They were against Chinese labour but they were more against 'interference in our affairs'.

The other main British colonial party was equally opposed to British government interference but were wary of being granted self-government so soon after the war. They were also in favour of Chinese labour. The Transvaal Progressive Association was set up in November 1904 in opposition to the Responsibles, and took their name from the Progressives in the Cape, since they planned to have similar views and membership. Instead of self-government, they advocated the creation of a largely nominated legislative council nominated by the High Commissioner, although their party manifesto also specified their opposition to 'interference in affairs of the Transvaal of party politicians elsewhere'.[46] Because the leader and deputy leaders of the party, Farrar and Fitzpatrick respectively, had such good relations with Milner, they only became particularly politically active from February 1905, when Milner left and when it looked certain that the Liberals would win the next British election. Then, the party's chief policies were their push to have a 'full investigation' before a decision was made about Chinese labour, and generally strengthening links with Britain while retaining democratic independence.[47] Unlike the Cape Progressives, however, their powerbase never managed to encompass the majority of the British colonial population as they had hoped. There was support, as in the Cape, from Britons scared of an Afrikaner majority.[48] Certainly, the Progressives saw it as their responsibility to advocate a form of self-government which would ensure they were not 'swamped at the polls'.[49] Much effort was also put into persuading the public of the necessity of Chinese labour. Sources sympathetic to the mines reiterated the economic dependence of the region on the mines and 'one-third of the entire industry is today absolutely dependent on Chinese labour.'[50] The president of the Wit Deep local branch of the Progressives claimed that the party did not want Chinese labour but thought it necessary. They could have 'flooded' the country with Southern Europeans but did not in order to protect British white interests. There was bound to be a bit of crime but better policing and understanding of their character would help, and in the mean time, they should keep the Chinese until an alternative was found. To further help recruit Rand labourers away from the other parties, they even supported the introduction of an eight-hour working day on the mines, which was very popular.[51] There was even gossip that Farrar and other magnates were willing to give extra voting rights to Afrikaners in return for keeping the Chinese.[52] Meanwhile, Sam Evans, a partner in Eckstein's, accused the British government of having 'certain socialist tendencies' in their interference with the labour scheme.[53]

However, the public perception that they were a party controlled by mining magnates, which left many feeling that 'none of them [were] to be trusted any further than they may be seen'.[54] Their portrayal as 'the capitalist party in the Transvaal' was not entirely fair.[55] Instead, their original membership consisted mainly of loyal Britons on the Rand involved directly in trade or mining, and worried about Afrikaner power.[56]

This left a great many British colonials, however, and yet another party formed in 1905, a small Transvaal Independent Labour Party. It was made up primarily of the 'labour aristocracy' on the Rand. Very few were from South Africa itself. While opposition to Chinese labour was their chief campaigning issue, the leaders – Creswell, Wybergh, and Outhwaite – also used their platform to promote 'white labour'.[57] Wybergh led most of their campaign in the Transvaal, while Creswell and Outhwaite spent most of their time in Britain.[58] Wybergh described the issue as 'a national question rather than a question of economics and mining costs... The mines belonged to the people of the country and they should see that they got their fair share'. In line with Creswell's white labour policies, 'the white men would substitute Chinese.' The party pushed for repatriation at the end of current contracts, with no renewals.[59] Broadly supportive newspapers like the *Transvaal Advertiser* in Pretoria went so far as to advocate a 'Back to the Land' policy as the only way to secure a national 'prosperous and happy' state.[60] Creswell's plan to get white unskilled labour was problematic, however, in a place where white men did not do 'kaffir work' and unionists feared such white labour would undercut their position. Furthermore, they frequently appealed to Britain to immediately end the scheme, which was why Creswell and Outhwaite both spent so much time campaigning there. This went against the mood of most Transvaalers; whether for or against Chinese labour, they did not want the decision to rest with an imperial government. The main Labour Party's inaugural meeting in 1905 attracted 100 people or less, out of the 15,500 Britons working on the mines.[61]

As for the Afrikaner population, a few meetings were held in 1903 and 1904 but they did not meet until January 1905 in Pretoria as the 'Organisation of the People' or Het Volk.[62] Most of the policymaking was dominated by Smuts, with Botha as the charismatic leader. To Smuts and to other leading Afrikaners, self-government, followed by South African federation, were the only ways to prevent the magnates dominating politics: even if the mines could dominate the Transvaal, they could not if confronted by 'the people of South Africa'.[63] They therefore advocated the 'promotion of closer union with other colonies' in southern Africa, an 'increased say of parents in the education of their children', 'protection of white labour' and no more Chinese immigration.[64] More than any British party, they were able to appeal beyond their 'racial' group, largely because they were trusted to be anti-capitalist and care about the future of the country more than the wealth of foreigners. Kruger had been re-elected in the ZAR repeatedly partly because of his anti-mine attitude. As early as the 1890s, the *Standard and Diggers News*, a pro-Kruger newspaper, advocated many of what would later be Creswell's policies about white labour, 'to run the mines in the interest of the white community at large rather than of a few Randlords'.[65] Few Afrikaners were involved in the mining industry and much of the Calvinist doctrine taught in churches celebrated agriculture and condemned modern capitalism.[66] 'To my mind the Chinese question is becoming the root question of Transvaal politics. If the magnates win there, they will rule this country as sure as Satan rules in the world below. But beaten there in their central contention, and there will be hope yet.'[67]

As in 1903–04, there were other Afrikaner views, of course. General J. G. Kemp, later leader of the Nationalist Party and part of the 1914 Afrikaner rebellion, supported Chinese labour, believing that it would secure more 'native' labour for farmers, who were predominantly Afrikaners.[68] Some farmers as well managed to make a decent living from selling pork and other food supplies to the mines, for use by the Chinese, an argument particularly common in the ORC. [69] Regardless, after the first official Het Volk meeting in January 1905, very few Afrikaners expressed public support for Chinese labour.

Such public unity was clearly strengthened by Chinese 'outrages'. Political activity increased and many public meetings were held to protest over poor policing, lack of compensation, and the general state of fear which persisted. Most requests for rearmament were organised by Het Volk. The size and regularity of the meetings alone indicates widespread Het Volk concern. Over 400 people met in Heidelberg to request 'respectfully' for the 'Government to afford better protection to its subjects'. They mentioned 'murderous outrages' and 'roving bands of Chinese', for which they held the mines and government jointly responsible.[70] Another meeting in Pretoria on 4 May unanimously passed a resolution claiming 'the Chinese are imported exclusively for the benefit of the mines and the country population suffers heavily'.[71]

Responsible or representative government?

The formation of these different parties was predicated on the long-standing British plan to eventually grant self-government to the Transvaal. It was the timing and nature of this which was unclear. In Britain, the Unionist government originally planned to extend the intermediary legislature appointed by Milner, and Alfred Lyttelton, the Secretary of State for the Colonies, accordingly drew up a report about the terms of this in April 1905. The Lyttelton Constitution outlined a nominated Executive with a smaller number of elected members. An electoral roll was drawn up and, until the Liberal victory in December 1905, everyone assumed that there would be an election on those terms in March 1906.[72] Another preconception was the widespread belief that there were more uitlander than Afrikaner men in the Transvaal; the government unofficially estimated there were 58,000 adult British men and 34,000 Afrikaners. However, the 1904 Transvaal census showed this might not be the case when officially published in mid-1905. There were far more Afrikaner women and children, and the number of men were quite close. The general assumption that the census was imperfect, given the continued post-war displacement, meant the exact number of eligible men was debatable.

The Progressives were against full self-government and favoured the Lyttelton constitution because it would protect against a possible Afrikaner majority, but the other political parties, and most colonials elsewhere, supported self-government, despite the risks. Once the Liberals were in power, they dismissed the Lyttelton Constitution as a viable option and instead appointed a Committee of Enquiry, led by Joseph West Ridgeway, to draw up a self-governing elected constitution of the Transvaal.[73] He allegedly admitted privately that his remit was to ensure an anti-Chinese majority in the first election.[74] In order to achieve this, the West Ridgeway Committee initially tried to arrange some sort of power-sharing agreement, with Richard Solomon as prime minister, so an election could be avoided.[75] All Transvaal parties were brought to London in March 1906 to negotiate over this. This would have given them more time to settle the final question of self-government and guaranteed a moderate British colonial government and against Chinese labour, but was unsuccessful. Several Progressives did not trust Solomon after his change of policy about Chinese labour[76] and there was little Transvaal appetite for such a compromise. There was also no agreement about how constituencies were to be drawn up or Chinese labour.[77] All parties instead accused each other of corruption, of rigging resolutions, petitions, and meeting votes, as well as massaging the number of white miners actually employed.[78]

The Committee eventually proposed to disallow British soldiers from voting, men who would have most likely voted for the Progressives. Likewise, the Lyttelton Constitution had called for economic criteria but this was removed, in order to ensure the wealthy mining magnates and their confederates did not have undue influence, and to avoid disadvantaging poor Afrikaners.[79] In exchange, the Committee persuaded the parties to agree to reject constituencies on the basis of population as Het Volk had wished (women and children to be included), and choose instead the Progressive and Responsibles-championed and Lyttelton-recommended 'one-vote-one-value' system, which only counted men in constituency population size. This decision in theory would discriminate against rural, largely Afrikaner populations and favour the more densely male-populated urban areas, chiefly Johannesburg. These choices, all adopted by the Liberals, reflected their desire for a small Het Volk or British but not Progressive majority. Using this as their basis, the parties all eventually agreed to the Progressives' suggestion that 34 seats be given to the Rand, 6 to Pretoria and 29 to the rest of the country. This was then adopted by the West Ridgeway Committee.[80]

The transvaal political Campaign, 1905–07

Broadly, the ensuing election campaign was divided between the Progressives and a loose collaboration of Het Volk, Responsibles, and Labour, united in their desire to limit the power of the mines.[81]

On 15 March 1905, the Responsibles had publicly announced that they would negotiate with Het Volk and by mid-April, an agreement had been reached to campaign together, although the bulk of the Responsibles were kept in ignorance until after the agreement was settled.[82] Likewise, at one meeting, a joint Het Volk and Labour resolution that no candidate would be supported unless against Chinese labour was carried unanimously, and there was further agreement that the parties would not oppose each other in most contested seats.[83] In order to strengthen the image of themselves as the primary party and once 'responsible government' was no longer in doubt, in September 1905 the Responsibles became the Nationalists and immediately reached another 'informal agreement' with Het Volk to support each other's candidates and not run against each other in several urban constituencies.

This cooperation offered several advantages. Clearly, to Het Volk, it allowed them to extend their support beyond the Afrikaner community and thus foster the cooperation of the Liberal government in Britain, while the Responsibles and Labour were given a prominence unreflective of their actual numerical support. When corresponding with Liberal MPs, CO officials, or the West Ridgeway Committee, Het Volk and the Responsibles deliberately presented the Responsibles as the leading party, with Het Volk merely supporting them. Smuts went so far as to state several times that, even with altered electoral criteria, Het Volk would only ever be able to achieve a minority.[84] In this way, the Responsibles were able to gain greater say in the Liberal government's policies than the small and fractious party really warranted.[85] As a consequence, the West Ridgeway Committee gave them twice as much space in the final report than the far larger Progressives, and clearly envisioned them winning the election, with a Het Volk minority.[86] The better-organised but small Labour party was also able to 'punch above its weight' by combining with Het Volk on the campaign trail. Additionally, most Labour unionists disliked the mining magnates more than Afrikaners; most members were recent arrivals from Australia or Britain so did not harbour the race hatred characteristic of other uitlanders.[87] Labour failed to unite with the Responsibles, however, because so many Responsibles were involved in the mining industry, although they did occasionally campaign together.[88]

When the election was finally held on 20 February 1907, Het Volk won 34 of the 69 seats, with 43 candidates.[89] They took all the country seats except Barberton (home of the mine with the second largest number of Chinese), three urban seats in Pretoria and four on the Rand. Three additional independents were aligned with Het Volk, giving them a clear majority. The Progressives put forward 34 people, 21 of whom won, 20 on the Rand, and 1 in Pretoria. All five mining magnates who ran in their mines' towns won, although again, accusations of intimidation and corruption overshadowed their results. None of the Progressive victories were as substantial as many Het Volk victories were, reflecting the more politically divided urban British populations where they won. The Nationalists put forward 16 candidates; 6 won, 4 in Johannesburg and 2 in Pretoria. The defeat of their leader, Richard Solomon by Fitzpatrick, was a particular embarrassment, but Het Volk appointed him Agent-General to London.[90] Labour put forward 14 candidates, winning three seats in Johannesburg, although their controversial leader, Creswell, lost by a small margin. Broadly, the results reflected a Transvaal political community divided by whether they supported Chinese labour or not, whether they trusted the mines with political power or not.

The labour problem revisited

After the elections, Botha was appointed Prime Minister and Minister of Agriculture, Smuts was Colonial Secretary, J. de Villiers Attorney-General and Minister of Mines (despite no substantial mining experience), Hull was Treasurer, Rissik was Minister of Lands and Native Affairs, and Harry Solomon became Minister of Public Works. Edward Rooth, a Pretoria attorney, became chief government whip in the Transvaal Assembly and a Nationalist, R. Goldmann, was made Assistant whip.[91] In opposition, the Progressives had Abe Bailey as Chief Whip and George Farrar as Leader. Fitzpatrick's election was a convenient excuse for easing him out of the partnership, as Phillips, Eckstein, and Wernher had long wanted to do.[92]

Initially relations between the two sides of the legislature were highly acrimonious. During the first session in March 1907, Chaplin tried to get consideration for a motion 'requesting' the government 'immediately to appoint a Commission to inquire into the question of the unskilled labour available', although without success.[93] Fitzpatrick, too, made an overture for compromise with Het Volk but was told Botha's party was not in favour of a labour commission. Het Volk thought African and white labour sufficient for all mining needs, but Botha apparently also promised Fitzpatrick that 'if it is not possible we are not going to ruin the Transvaal by turning out the labour that is here now' and promised not to 'act in a hurry'.[94] Smuts in particular had reluctantly (and certainly not publicly) accepted the dependence of the state's finances on the mines, although he still hated the thought of them wielding political

Table 7.1 Repatriations of Chinese labourers at the end of their contracts, 1907–10

	1907	1908	1909	1910
Maximum number at work in January each year	53,838	31,480	10,045	1907
Total loss during year	22,348	21,435	8138	1907
% loss during year	41.51	68.09	81.01	100
% of unskilled labour employed at 30 December each year	23.77	7.12	1.01	.81

Source: Annual Report, Chamber of Mines, 1907–10.

power.[95] Eventually they compromised with a policy of 'replacement'[96] rather than forced repatriation immediately. By staggering the removal of Chinese labourers until the natural expiry of their contracts, Het Volk could ensure that their victory was not accompanied by an even worse economic depression. Likewise, by ensuring that all Chinese would be repatriated by January 1910, there was time to establish labour replacements and to point out to the strongest anti-Chinese voices that the scheme had an end date (Table 7.1).

Once it was decided to phase out Chinese labour, initially conditions on the mines changed very little. Indeed, there was a small petition sent from some Chinese labourers to object to the cancellation of the scheme, and there was a brief increase in deserters caught trying to reach Mozambique, but for unknown reasons. The press continued to report any disturbances, although the matter was now largely voyeuristic rather than an active panic. From April 1907, the government even stopped centrally recording the rates of mortality or crime among the Chinese, since they no longer had to justify the scheme to the general public.[97] Violence as a form of supervisory management continued, now largely unchecked, and would continue to plague mine culture for decades to come.[98]

Instead, the labour problem once more became a political issue in the Transvaal. If the mines were to survive as viable investments, Africans would have to be recruited in far greater numbers than ever before, and stay for longer periods, to match the efficiency and numbers of the Chinese.[99] Most viewed the recruitment of Africans as unpredictable, and did not relish having 'to subject the vast interests represented here to the caprice of the native'.[100] Despite concerns, however, it soon became apparent that Chinese labour was easily being replaced by Africans (see Table 4.3 and Table 7.1). This was aided by several factors. Many contemporaries thought 'the Kaffir has learnt a salutary lesson through the importation of Chinese coolies. He has seen that he has to seek work in order to obtain it'.[101] Others blamed it on African racial characteristics: 'With the perverseness of children, the natives, seeing that they were no longer indispensible, began to offer their services to the white man with increasing readiness.'[102] More realistically, public works and private expenditure had all been curtailed during the heightened depression in 1906 so there was less competition for African labour than there had been immediately after the war.[103] Prolonged drought and a diamond recession further increased the influx of natives, particularly from the Cape. From 1909, there was swift economic recovery to the mines and farms in southern Africa, but great displacement of whites and blacks in rural areas, most of whom turned to mine-related industry for employment.[104] WNLA recruitment in 1905 had actually been 39 per cent higher for natives than in 1904; successive years saw this trend continue, suggesting Chinese labour had never been necessary.[105] The mines were also in a better financial position in 1907 than they were in 1903, despite their poorly performing shares and overcapitalisation. In 1903, mine output was only £12,500,000, whereas in 1906, the output was £24 million.[106] Some of this money was being invested in providing better living accommodation, modelled on that set up for the Chinese, and this made work on the mines marginally more attractive.

Most importantly, the government helped negotiate recruiting from the Cape and Mozambique. Between 1903 and 1906, 154,047 Mozambiquans were recruited through official avenues, staying usually for one-year contracts.[107] While this was much less efficient than the three year contracts the Chinese had, this was increasingly made up for in other ways. African recruiting cost £10 15s per head and Chinese £15 11s, leading to immediate savings. Furthermore, because of the increasingly large number of regulations governing the administration of the Chinese, the cost of maintaining them was almost double maintaining natives.[108] There was significant expense, and loss, in closing recruiting centres in China,[109] but the cheaper African labour helped this transition. Furthermore, the mines began rapidly to utilise more efficient mining techniques recently developed in America, especially rock drills. They also moved white supervisors from overseeing 3–4 Chinese to 7–8 on most mines. Both of these measures theoretically meant that less labour was actually needed to do the same amount of work.[110] While in 1903, it had been widely considered that African labourers needed more supervision than this, mines were reluctant to go back to the old ratio, so the system remained in place, but with Africans in place of the Chinese.

Ideas about government intervention to solve social problems had been increasing in popularity for the past 20 years in Britain, largely through the activities of the Fabians and New Liberals, as well as the growing labour movements. Likewise, in Australia, the White Labour Policy relied upon assumptions that the government could and should be used to fix social problems.[111] Such ideas had proved hugely attractive to the politically active white populations throughout southern Africa. Milner and his kindergarten had always been in favour of government intervention to shape South Africa into the colony they desired. The implementation of Chinese labour was a reflection of their dedication to utilising a central government to shape society. Subsequent reactions to Chinese labour in the Transvaal and the links which developed with white labourism in Britain, Australia, and elsewhere yet more firmly entrenched such ideas into the minds of South Africans themselves. While people debated whether Chinese labour was desirable or not, the central principle that the government should use its position to shape society went increasingly unchallenged, as long as it was the local democratically elected government and not an ignorant imperial government doing the shaping.

Het Volk did little to change this evolution in the role of the government in South African affairs. The scheme had established a precedent for intense government oversight and Het Volk not only refused to relinquish this power, they strengthened it. While they negotiated new recruiting drives for the mines, Het Volk also created a Native Labour Bureau. This would supervise native recruitment and living conditions, much the way the FLD had done for the Chinese. No such organisation had hitherto existed in British Africa to supervise recruiting; it came directly out of the experience with Chinese indentured labour, although historians have incorrectly compared it to the Rhodesian Native Labour Bureau, not the FLD.[112] The organisation was also established because, in order to get rid of Chinese labour, the new government wanted to prove that sufficient other labour sources existed. In the short term, the extra money thus supplied to supervise WNLA recruiting activities, and the combined desire of the Liberal and Het Volk governments for sufficient African labour to be found, meant that WNLA had greater government support, although had to function very much on the government's terms.

This was most noticeable when it came to the matter of unskilled white labour. While the upper echelons of Het Volk were doubtful of the willingness of white men to do 'kaffir' work, they were in close alliance with the Labour Party. There were also a high number of unemployed Afrikaners, most of whom voted for Het Volk, if they voted at all. Consequently, in September 1905, Het Volk had first approached the Chamber of Mines secretly to ask them to consider placing poor Afrikaners on the mines at 5s a day.[113] While more than Africans were paid, the mines agreed to try this in 1907, largely through the support of Lionel Phillips of Corner House. In return, Het Volk had promised to allow the Chinese to finish their contracts before repatriation and to assist in recruiting.

This development was not without opponents, however, and led to an uncomfortable partnership between the government and mines against protesting skilled white miners, most of whom were British. During the famous May–July 1907 strike, the two major complaints of strikers were the use of white unskilled Afrikaners and the changed ratio between white supervisors and their non-white 'boys'. Membership of the Transvaal Miner's Association increased dramatically as a result and the labour movement generally was given a huge boost.[114] Despite this, the mines gained the backing of the government, largely by replacing the men with Afrikaners. The proportion of South African-born white mine workers increased subsequently from 17.5 per cent before the strike to 24.6 per cent after. There was a slight decrease after the strike, but then the number steadily increased throughout the twentieth century, except in 1922 when another major strike occurred.[115] It was exactly this shift from employing imported artisans to employing and training southern Africans which ensured the government's cooperation and fostered better relations among the two groups. Even after the strike, Corner House continued experiments with white unskilled labour, working closely with Het Volk's Mining Minister.[116] In the short term, this cooperation clearly paid off. By September 1907, many mines 'exceeded all previous output'.[117] Leopold Albu, during one speech, even optimistically declared that 'not a single Chinaman had been repatriated without a substitute being found' (see Table 4.3).[118]

When examining the post-election 'labour problem', most historians have focused either on the increased unionisation of white labourers or on how the mines were able to permanently lower African wages because the Chinese undermined their ability to demand higher wages. Indeed, labour union membership did increase while African wages after 1907, in real terms, declined, when the wages paid to Chinese labourers were forced on them.[119] What have been less well studied are the changing racial views at this time. As mines increasingly utilised Africans in the same way as they had Chinese, the idea that Africans were not capable of competing with whites began to die away. The ratio between white and Chinese workers, since their advent, had decreased from 19 per cent to 11 per cent.[120] Once Africans started being reintroduced to the mines in 1905, the same cost-cutting measures were used for their supervision. The success of this 'indicated to management that there was nothing inherent in the African which might "prevent" him from doing work of a similar nature to that of whites'.[121] Indeed, a government report in 1908 celebrated their capacities as workers.[122] This vulnerability to replacement by cheaper, less unionised, African workers made many skilled white miners increasingly eager to impose the colour bar of the Chinese ordinance against Africans and more likely in future to project economic and political anxieties onto Africans in the form of Black Peril panics.

In order to appease this growing number of labour activists, Botha's national government passed the Mines and Works Regulations of 1911, which helped secure skilled labour purely for whites, by adding to the list of restricted jobs, first drawn up in relation to the labour importation ordinance of 1904. This was despite the protests of the Chamber, who wanted to continue to use African labour in such positions. Increasingly Africans were 'considered intruders in the "proper sphere of the white man"'.[123] Regulating and controlling their lives was not new, but the increasing desire to consider them potential competition and the increasing level of laws and tighter enforcement owed much to the legacy of the Chinese administration and the relationships which had been established between the different bodies involved in the scheme.

As a result of this, and the governments' support for mines during the 1907 strike, increasing numbers of skilled miners turned to Creswell's version of 'white labourism' and to his Labour Party. For the next election in 1910, a nation-wide South African Labour Party, still headed by Creswell (largely thanks to the fame of his anti-Chinese campaigning and pamphlets in southern Africa), won five seats. This time, Creswell himself won a seat, and remained in parliament as head of the party until 1933, directing a white labour policy strongly underpinned by his own vitriolic feelings towards the entire Chinese labour scheme.[124] At the 1914 election, Labour was even more successful. This led in turn to a deal with the National Party in 1920 since both now saw the South Africa Party, led by Botha, as complicit in a suppression of worker rights. Labour won 21 and the NP 44 out of the 134 seats.[125] This partnership was strengthened by the 1922 strike[126] and they went into formal government partnership from 1924.

South African Union

This was not the end of Chinese labour as a political issue, of course. When the Transvaal and ORC were granted responsible government in 1906 and 1907, respectively, the talks of South African federation were 'renewed' and, despite its cancellation, Chinese labour continued to pervade many aspects of discussion. While advocates of Chinese labour importation had once promised that it would secure the economic and *British* future of all of South Africa, now they increasingly argued that union was the solution. To both Progressive parties and remaining members of Milner's kindergarten, union would secure the nation for Britain and would help push forward imperial federation. Even colonials who were not particularly imperialistic placed their economic hopes now in union.[127]

A separate group also organised to champion union, albeit largely out of suspicion of mine power. Merriman, Steyn, Smuts, F. Malan, and Botha in particular had fostered a strong network of ideological exchanges during their opposition to Chinese labour. All of them were united by their desire to ensure South Africa was run by *real* South Africans, not outside groups. In their views, while the gold mines were economically important to the future nation, most mining magnates were decidedly not South African. Their philosophy, born out during the 1904 Cape and 1907 Transvaal elections and fostered by a shared opposition to Chinese labour, was that farmers and urban workers, Britons and Boers, should join together because 'it was to the advantage of both to join forces against the capitalists'.[128] As Merriman told Steyn: 'I want to do something to take the settlement of South African affairs out of the hands of Downing Street, and at the same time to make us feel that we are members of one body – that the franchise and Chinese questions are South African and not local affairs'.[129]

Smuts agreed: 'unless the power of the magnates in the Transvaal is broken by our entry into a unified or federal South Africa, the danger of their capturing supreme power here and so over the rest of South Africa...will continue to exist'.[130] Fischer, the Natal premier, Merriman and Steyn even considered 'that some union between the ORC and the [Cape] Colony will be the only chance to set up some sort of barrier against the demoralization of the whole of South Africa on the Jo'burg model'.[131] They often mentioned the Chinese scheme as *the* example of what terrible things could happen if the mines were too powerful in the future union.[132] When Selborne tried to negotiate an unelected coalition government for all South Africa made of all the different parties, these men all vetoed the proposal because 'such a project would...place us virtually under the control of the money power'.[133] In addition to anti-capitalist feelings, these sentiments played on a widespread fear in southern Africa that the Transvaal itself would dominate national affairs. 'While the idea of union is outside of Cape Town, more popular than it was, in the abstract, the discussion has shown in many quarters a strong and growing feeling of distrust and of apprehension at the prospect of being subject to Transvaal hegemony.'[134] Indeed, it was because of this concern that they also vetoed the use of Johannesburg as a national capital, and demanded that the seat of government be split between Cape Town and Pretoria, as it eventually was.[135] They also concluded that: 'The only way...out of our political and economic difficulties is a union not a Federation of South Africa.'[136] Marthinus T. Steyn in the ORC viewed the conflict between Britain and the South African republics as a 'struggle between capitalism and individuality, between imperialism and, on the other side, republicanism and colonialism'. He embraced racial (Britain and Boer) harmony and celebrated farming, not industry,[137] even suggesting on one occasion that the ORC could be annexed to the Transvaal, in order to stem the power of urban capitalists there.[138]

Meanwhile, many imperialist-minded southern Africans were concerned that the 'country' parties of Het Volk, the Bond, the SAP, and Orangie Unie would unite in a federal government 'and won't be sympathetic to mining interests' or to the British Empire, which, as in 1903, they argued depended on each other.[139] Several members of Milner's kindergarten, while less closely aligned with the mines than he had been, worked with leading Progressives from the Cape and the Transvaal to launch the Closer Union Society in January 1907 in both colonies, while ignoring the ORC and Natal altogether.[140] By 'encouraging, local colonial nationalist sentiment, and by thinking of "British" identity in terms that allowed for overlapping loyalties',[141] they sought to address the accusations so often made that they were not *real* South Africans. Such a discourse closely mirrored the arguments they had utilised to justify Chinese labour. Consequently, it is hardly surprising that the two political groups most closely involved in mapping out the terms of union were divided in a way that also mirrored the political divisions and partnerships regarding Chinese labour. At the same time, despite the occasionally acrimonious relations between them, matters progressed surprisingly quickly, partly because many of the debates had already been played out through the Chinese labour controversy. This explains in part why it took ten years for Australians to agree on the terms of federation but less than three for South Africans. Official negotiations about the terms of union began in June 1908, when the Cape parliament petitioned Selborne to grant union. A National Convention was held from October 1908–February 1909, largely masterminded by Lionel Curtis, Smuts, and Merriman. The British Government passed the South Africa Act in 1909, approving the constitution. While they no longer were willing to interfere outright in South African race politics, after the Chinese political problems, they did keep Basutoland, Bechuanaland, and Swaziland out of the union, to protect non-whites there from southern African racial prejudices.[142]

Indeed, the wish to keep the British from interfering in their affairs had been an important shared concern among delegates from all parties. Farrar, when corresponding with Selborne about the draft of the constitution, wanted to make sure that they did not 'give HMG a chance to interfere in future',[143] a view vehemently shared by Merriman, Steyn, and Smuts.[144] Chinese labour again was much cited as *the* example of why British interference was so decidedly unwanted. A British government had approved the scheme, against the wishes of many, and a different British government had disallowed it, despite objections. No one wanted the British government to have the power to so dictate colonial affairs in future; they did not want their colony to be ruled by political squabbles or whims 6000 miles away. Union was seen as a way to prevent this, even among loyal imperialists.

Union was also affected by Chinese labourers in less obvious ways. One of the major reasons for union was to facilitate the exchange of goods and workers: if the colonies had to compete for labour, this would not 'conduce towards early federation'.[145] The Transvaal's new agreement with Mozambique, drawn up in 1909 to provide additional labour for the mines through WNLA, was a particularly strong argument to Natal to not enter the Union, since it promised Delagoa Bay at least 50 per cent of their overseas shipping for the next decade. While the Chinese labour importation had been profitable to the colony, future migration would largely bypass them.[146] However, the bay in Durban has been extensively expanded, while storage and transport facilities greatly improved from 1903, largely at the expense of the mines. While the transportation of labour through the Durban would cease, the bay was now secured as the largest and most developed port throughout southern Africa; Delegoa Bay, now Lorenço Marques, would never seriously compete again.[147] Furthermore, many Natalians were worried about Afrikaner dominance of the Union. Natal's fears of African and

Asian dominance, heightened by the 1906 Zulu Rebellion and Chinese unrest, were successfully played upon by all sides advocating union, and Botha's conciliatory measures regarding the transfer away from the Chinese further assured them.[148] At the 1909 union convention, Botha said that the gold mining industry and all South Africa needed a new treaty, 'especially since the Chinese labourers were being repatriated and the industry would soon be wholly dependent on African supplies of unskilled labour'.[149]

Conclusion

It is clear that Chinese labour had a major impact on the politics of Britain and southern Africa. Analysing the political discussions in 1905 onwards also shows how ideas of white labourism, democracy, imperial federation, Britishness, capitalism, and the role of the state in shaping society were reconfigured by the Chinese scheme. Imperial federation was no longer of widespread concern, although national federation in South Africa was. Afrikaners had gained political power and largely South African nationhood was built on the unity of the two white races under Britain, as long as British did not interfere too much again. The British government had learned that interference in settler colony affairs was politically dangerous and perhaps best avoided, even if it meant subsequently allowing race-based legislation.

Asians were also almost entirely left out of the Union, both within the constitution and in the accompanying pageantry and commentary, largely reflecting the South African view that Africans were *native*, and that Britons and Boers could be *real* South Africans too.[150] The popular increase in anti-Asiatic sentiments had been fostered by a sense that the imported Chinese labourers had been alien, sent to replace the legitimate *South African* workers on the mines. Much of the negative feeling assumed Asians did not belong in South Africa, that they were neither native nor legitimate immigrants into the nation; furthermore, they were incapable of assimilating, so could not become South African. Davenport has even noted that the Het Volk manifesto, written primarily by Smuts, but with input from Botha, Malan, de Waal, Merriman, and others, 'offered something general to all major ethnic and interest groups save the Asian'.[151] While he does not expand, it seems likely that Asians were left out precisely because of these feelings that Asians simply did not belong in South Africa and could never be South Africans. After 1907, fresh Asian migrants were almost entirely prevented from migrating to southern Africa, a position not altered until the 1990s.[152] This was fed by reports of Chinese unrest, which continued sporadically until the final repatriations in 1910, and repeated racialised portrayals of the Chinese, fostered by the numerous memoirs published later, many of which glorified in the most salacious aspects of the scheme.[153] That became the enduring memory of the scheme and the enduring twentieth-century depiction of Chinese peoples in South Africa.[154] Indeed, there continues to be a myth in South Africa that the permanent Chinese population originated from Chinese indentured labourers who managed to escape. Officially in the FLD files, only 12 Chinese went untraced at the end of repatriation, while local historians have traced only 2 possible escapees. Nevertheless, the story has persisted.[155]

Clearly then Chinese labour had a lasting and profound effect on South Africa. Saul Dubow has correctly noted that 'one of the major themes of this period – the ideological construction of white "South Africanism" – has been pushed to the margins, and the effusion of political and cultural activities associated with the creative imagining of the first "New South Africa" has been overlooked, or mentioned only in passing'.[156] This study has provided an analysis of one of the most significant aspects of the 'construction' of that identity and thus enables southern African historians to piece together the formation of the nation. The reconstruction period should not merely be marked for being a period 'when South Africa might conceivably have taken an alternative road toward a more liberal society' and where 'the Milner regime established a close alliance with the mining magnates and made the crucial decision to consolidate and defend a cheap-labor policy'.[157] Such statements fail to understand the complexities of the reconstruction period and the effects of the Chinese labour controversy on a global scale.

Conclusion: Racialising Empire

Asian migration controversy

Perhaps the most important long-term legacy of the Chinese scheme was the cementing of a two-tier British Empire. This was most clear in the way colonial and imperial governments interacted over race-based legislation after the scheme. Before it, as Chapter 1 showed, Britain banned such laws, but because of the Chinese controversy, things changed. It was hoped that successive CO regimes would clarify the matter. As late as March 1906, the *Indian Opinion* was declaring that

> we coloured people have much to be thankful for to the Celestial labourers, who have...become the lever by which the late reactionary Government was turned out of power. Chinese labour may yet prove to be the salvation of the British Indians in South Africa, if it be the means of arousing the conscience of Britain to a sense of its enormous responsibility in regard to the treatment of the coloured races.[1]

The free Chinese also hoped for greater rights under a British government, as opposed to the racialist policies of the ZAR.[2]

Much to their dismay, the willingness of a Liberal-run CO to proclaim on the matter was minimal after the hassle over Chinese indentured labour.[3] Partly, this was because most British parliamentarians could not follow 'the differences between the ordinary Asiatic, British Indians, Natal indentured coolies and the coloured people of the Cape'.[4] After the controversy over Chinese labour, there was an implicit acceptance that the imperial government did not have the knowledge or power to decide such matters. When the Transvaal parliament passed the popular

Asiatic Law Amendment Act, the British government refused to overturn it.[5] The same was true when the first act of the new parliament in 1911 was to ban Asian migration throughout the country, and Natal finally stopped indentured importation in the same year.[6]

Nor did this matter only affect South Africa. By 1908, even with the scheme cancelled, newspapers could still write 'that every part of the Empire is interested in the situation presented in the Transvaal with regard to this... problem of Asiatic immigration'.[7] The reason this continued to be so prominent was because other settler colonies still wanted to restrict Asian migration, whether British subjects or not, and the British government had still not allowed this.

If Britain let the Transvaal pass such legislation, the other settler colonies wanted to do the same.[8] Many Australians and New Zealanders advocated introducing something similar in their colony, which the imperial government would no longer be able to veto.[9] When some pointed out that British Indians would be affected detrimentally, it became increasingly common to argue that it was 'entirely the right of autonomous British possessions to make their own laws, and to administer sanctioned laws without Imperial interference or modification after such laws have been assented to by the Crown'.[10] When the American naval fleet was sent to Australia, New Zealand, and Canada in 1907, against British wishes, the colonials there referenced the imperial government's role in 'horrors of the Rand' as a reason for promoting American, rather than British, protection against the 'yellow peril'.[11] Various organisations who advocated restricting Asian migration in New Zealand used the alleged high crime rate of the Chinese in the Transvaal as evidence to support restrictions.[12] The scheme even became mythologised in labour histories, particularly in South Africa and Britain.[13] Referencing a past event like the scheme was a far easier way to foster a sense of shared identities and cultures than referencing the complexities of contemporary localised problems. Whiteness was often paramount, more so than Britishness.

Generally, the lessons of Chinese labour were taken to be that it was best for the imperial government to avoid interference in 'white colony' affairs.[14] The British press and public also became more reconciled to the ideas of immigration exclusion and the Asian 'menace'. Once the idea of Chinese immigration had been raised there in 1905, at the same time Britain passed its own immigration controls. Besides, the incident of Chinese indentured labour into the Transvaal had 'proved' how detrimental Asian migration was for white nations.[15] The widespread acceptance in Britain for the first time that Asian migration and British imperial rule could not coexist was an important shift in attitude. For the first time, Britons felt themselves to be threatened too. Those who did not feel endangered as such felt the entire use of indentured labour tainted Britain's humanitarian credentials, already damaged after the controversial South African War.[16]

Such a shift in British perspectives had a major effect on empire. In 1905, Richard Jebb's seminal *Colonial Nationalism* had warned that the British government's actions concerning Chinese labour might not just have consequences in South Africa but in all the white colonies and their debates with Britain about imperial federation.[17] Imperial federation would never be attractive if the metropole meddled too much in colonial affairs. Already by 1906, the CO had decided to focus on colonial, later imperial, conferences rather than pushing for imperial federation. In 1907 the settler colonies were given their own sub-department within the CO and were renamed the Dominions (at Laurier's suggestion), to signify their equality with Britain. Soon after the 1907 Colonial Conference, there was a CO investigation into Asian migration into the self-governing colonies, which concluded that the introduction of the Chinese into the Transvaal had strengthened 'the bias against coloured immigration in the self-governing Dominions' and 'if we do not take the initiative, the United States may stand out on and through this question as the leaders of the English-speaking peoples in the Pacific as against the coloured races'.[18] Ramsay MacDonald claimed: 'That one act [Transvaal indentured labour] did more to destroy Imperial affection and pride in Colonies where the colour repulsion is felt, than anything that has happened for many years.'[19] The threat of losing the Dominions, and their military and economic potential, and the belief in Britain that the colonies actually would cede and join with the United States if thwarted in their exclusionary aims, castrated future British governments. The writer of the report, Sir Charles Lucas, was put in charge of the new Dominions Department (he had previously headed the Colonial Office's West Indian and Eastern Departments), where he continued to reiterate this view: 'If Britain were to confront the Dominions over the race question, they might break away and form an alliance with the United States, creating a new political organisation, "having its roots in race affinity", that would be directly opposed to the idea of imperial citizenship, which took no account of race.'[20]

This belief became the bedrock of future imperial policy; Britishness remained the primary identity for many settlers but could not be taken for granted. The notion of imperial citizenship thus lost all meaning, despite the CO continuing to advocate it and to insist that Indians and other non-whites were not second-class imperial subjects. This policy was clearly at the expense of India, as Lord Curzon, a former Viceroy of India, Milner and their followers complained, 'for in raising the profile of the Dominions, they also marginalized India and Britain's other tropical colonies in imperial debate'.[21] Indeed, despite repeated Indian requests, they were not named one of the Dominions.[22] The 1911 Imperial Conference (renamed to sound more like a meeting of equals) unsurprisingly failed to define imperial naturalisation, nor could the Dominions agree about whether a uniform immigration policy was needed. The Dominions were happy to embrace partnership with Britain to an extent, and could envision working with each other to get this, but not to embrace the rest of the empire.[23] The British government chose to accept this and appease India in smaller, usually less successful, ways. As William Wybergh explained, the settler colonies, like South Africa,

> look for an Imperialism which does not seek to mould the various colonies and races of the Empire upon the British model... but which seeks to develop all that is best and most characteristic in each colony on the lines most suitable to its own advance... Thus we expect that the Transvaal shall be regarded, not as a gold or a dividend producing centre, but as a home within the Empire for the British and Dutch races, as the cradle of a nation, and of a nation which is neither Dutch nor British, but South African.[24]

It was unfortunate for Britain that one of the greatest uniting factors among its settler colonies, by focusing primarily on race, undermined the British Empire itself. When the colonies and Britain focused on the racial qualities of Britishness or whiteness as the bond which tied them together, the links of empire had to give way. While the 'white' world was united through racial ties, the British Empire and the concept of imperial citizenship or British subject hood were devalued. This shift did not necessarily separate the colonies from Britain, but it did mean that colonies increasingly came to see themselves as separate from the empire. They were happier to rely on the mythical bonds of race to retain relations with each other and Britain and to avoid economic or political ties. This also allowed more flexibility to incorporate the United States within a wider English-speaking 'white' world.

Racialising empire

This specific moment in time: the end of the South African War; the widespread interest, yellow and black perils; the recent creation of Reuters and the Associated Press and their cable lines; rising literacy; uncertainty over whether Transvaal could ever be a white colony and over the place of capitalism within society; whether immigration should be controlled; to what extent the government should try to shape society; the imperial federation issue; uncertainty over who would have political rights in southern Africa, and specifically the position of nonwhites; debates over the position of British Indians within empire (and race based legislation); the rights of settler colonies versus the right of Britain; the spread of labour unions and rise of labour parties in politics; the spread of democracy; the ability to spread cheap images; and the last years of the Qing dynasty in China all show how useful it can be to examine a small scheme over a short period. This offers a crucial snapshot, or rather a series of snapshots, which show the nuanced interconnectedness of issues in different parts of the world throughout the period.

Networks had formed and changed as a result of the scheme. Various 'experts' had debated about the 'Chinese character' and shaped many lives based on their assumptions of race and class. Ideas about the Transvaal 'labour problem' were reconfigured, with greater government involvement, and the positions of 'white' and 'black' and 'yellow' labour in southern Africa were profoundly affected. The continued debates over who had a say in Transvaal affairs affected elections in southern Africa and Britain and shaped transcolonial relations. There was a transcolonial flirtation with imperial federation, and then a disillusionment, resulting in the formation of the Dominions and an increasingly racialised identification as 'white', instead of British colonies.

The analysis of the scheme has also revealed much about the press at the time. It reveals how one of the first truly global news stories was portrayed, giving a richer understanding of the pull between the local and the global. It also shows how newspapers were used politically at the time, their role in spreading ideas and fostering networks. The global spectacle was built on newspaper coverage, although it is worth remembering the press role was marginal in the Transvaal moral panic.

Perhaps one of the most significant legacies related to the alleged link between the South African War and British economic imperialism, a link which came to dominate historiographical examinations of both the War and the reconstruction period. Accusations of corruption did not stop the importation of the Chinese in 1904, but the whole controversy did seem to 'prove' to people the close relationship between Britain and the mines. As this book demonstrated, this was far too simplistic an interpretation of affairs. The relationship between the mines and imperial and colonial governments were fraught with tension. Indeed, the legacy the Chinese were supposed to leave was economic stability in the Transvaal, but this was not the case. When founded, the CMLIA was given access to up to £1 million to oversee recruiting. Instead, according to the Chamber, the experiment cost them £1,013,000, or £16 per Chinese labourer, a third more than the average cost for African labourers over the same three-year period. Bagot had even estimated they cost over £21 per head, and Richardson has calculated that, if the cost of the compound infrastructure built for the Chinese is included, the price was almost £32, most 'unrecoverable'.[25] This great expense and the detailed regulations in need of implementation were simply not compatible with fostering a harmonious relationship between those involved. Despite this, the public perception of a corrupt cosy relationship was lasting; indeed, Hobson's work has often been given credit for spreading the idea.

The scheme also affected empire significantly. The situation had revealed deep divisions among the settler colonies and Britain but it also better integrated some Britons within a shared 'white labour' identity, threatened by Asian migration. Opponents in South Africa, British Liberals, and labour organisations had utilised the issue for their own political ends. The Cape, Australia, New Zealand, and British Columbia had all attempted to use the principles of imperial federation to prevent Chinese labour importation, without success. It also makes clear that debates about the political relationship within this 'Greater Britain' were not limited to a colonial elite, or even to British settlers, but encompassed a diverse range of peoples and networks. There would never be such a widespread attempt to change imperial policy by the settler colonies again.

Appendix: List of Key Figures

Dr. Abdullah Abdurahman (1872–1940) b. Wellington, the Cape. Graduated from the University of Glasgow Medical School in 1893 and practised medicine before becoming the first South African non-white politician in 1904, sitting on the Cape Town City Council till 1940, and founder of the African People's Organisation (APO).

Sir Abe Bailey (1864–1940) b. Cradock, the Cape. Educated in England; arrived on the Rand in 1887 as a stockbroker and independent speculator. He was imprisoned after the Jameson Raid; was an intelligence officer in the South African War, but captured by the Boers. He served in the Cape and Transvaal Parliaments as Progressive in the Cape and an independent in Krugersdorp from 1907. He supported indentured labour importation but was also a leader in Asian exclusion. KCMG, 1911; Union parliament, 1915–24; baronet 1919, and largely made South Africa his home.

Alfred Beit (1853–1906), b. Hamburg. He arrived in Kimberley in 1873, becoming a leading diamond expert. Became partner with Wernher in 1880, co-founding Wernher, Beit & Co. in London and H. Eckstein & Co. in Johannesburg in 1889. A life governor of De Beers, though resigned from the board of the British South Africa Company for his involvement in the Jameson Raid. He lived in London.

Louis Botha (1862–1919) b. Greytown, Natal. A leading Boer general during the South African War, he went on to co-found Het Volk and be first prime minister of the Transvaal (1907) and first prime minister of the Union of South Africa (1910), as head of the South African Party.

Joseph Chamberlain (1836–1919) b. London, a manufacturer and radical MP for Birmingham from 1876. He served under the Tories as Secretary of State for the Colonies, 1895–1903. Resigned before approving Chinese labour to campaign full time for tariff reform in 1903. One of his final words was to accuse the government of conniving with the mining magnates over Chinese labour.

Francis 'Drummond' Percy Chaplin (1866–1933) b. Twickenham. Went to Harrow and University College, Oxford, then became a barrister at Lincoln's Inn in 1891. He was the Joint Manager of Consolidated Gold Fields of South Africa, Ltd (1900–14), President, Transvaal Chamber of Mines (1905–06) and a frequent Johannesburg correspondent for the London *Times*.

Alfred Child (?–?). Child was the highest-placed Chinese 'expert' among the CMLIA staff in South Africa and the most proficient British member of staff in various Chinese dialects, having spent 21 years in China; 17 were with the British Maritime Customs.

Winston Churchill (1874–1965) b. Blenheim. He had acted as a *Morning Leader* correspondent during the South African War, then won a seat in Parliament in 1900. He served as the Liberal Under-Secretary of State for the Colonies from late 1905 and oversaw the dismantling of the Chinese indentured scheme, defending government policy in the Commons.

192 Appendix: List of Key Figures

Frederick Creswell (1886–1948) b. Gibralter. A mining engineer, trained at the Royal School of Mines, he worked in Venezuela, Armenia, then the Rand (1893–1903), serving in the South African War. In 1902–03, as manager of the Village Main Reef mine, he headed the largest white unskilled labour experiment. Was controversially dismissed by Corner House, and then campaigned against Chinese labour in southern Africa and Britain. From 1905, he was leader of the South African Labour Party and served in the Union Parliament from 1910; imprisoned in 1913; served as Lieutenant-Colonel in Great War; Minister of Labour under Hertzog in 1924–25 and 1929–33.

Sir Thomas Cullinan (1862–1936) b. Elandpost, Cape. Went to the Rand in 1884 and began his own brick and tile works at Olifantsfontein; found fame with his mine prospecting (the Cullinan diamond was found in 1905). He was elected to North West Pretoria in 1907 as one of the few mining magnates who stood as a Het Volk candidate, winning again in 1910.

Lionel George Curtis (1872–1955) b. Derby. After attending New College, Oxford, he became the Private Secretary to the Rt. Hon. Leonard Courtney MP (1896–98) before fighting in the South African War, then becoming one of 'Milner's Kindergarten' as Town Clerk, Johannesburg (1901–03), then Assistant Colonial Secretary, Transvaal (1903–06). Important organiser and planner behind the Union of South Africa and later a campaigner of imperial federation. Also instrumental in pushing through the fingerprinting and registration of all Asians in the Transvaal.

Hermann Eckstein (1847–93) b. Hohenheim, Germany. He went to South Africa in 1882, becoming manager of the Phoenix Diamond Mining Company at Dutoitspan. He founded H. Eckstein & Co. on the Rand in 1887. From 1892, he lived in London, with his brother, Sir Friedrich, overseeing Johannesburg business. Hermann was also a partner in Wernher, Beit & Co. (1902–10); chairman of the successor, Central Mining Investment Corporation (1912–14). Baronet 1929.

John Emrys Evans, CMG (1853–1931) b. Northern Wales. A banker, he went to South Africa in 1882 as an inspector for the Standard Bank of South Africa, Ltd. In 1897, he became British Vice-Consul for the Witwatersrand Gold Fields, before becoming a financial advisor to Milner's administration and to Lord Roberts during the war, then Controller of Transvaal Treasury, August 1900, Auditor General, Transvaal 1901, Vice-Chairman of the National Bank of South Africa and director of several companies. Created CMG in 1902. From 1902, he was a leading Johannesburg financial advisor, sat on the Johannesburg Hospital Board, served on the Johannesburg Town Council for two terms from 1902 to 1906. In 1907 he was elected Progressive member of the Transvaal legislature for Langlaagte. He was Vice-Chairman of the National Bank of South Africa Ltd (to become Barclays Bank) and a Director of the Johannesburg Consolidated Investment Company. Brother of Samuel.

Samuel Evans (1859–1935) b. Northern Wales. A civil servant in Egypt from 1883, he later became Chairman of Glynn's Lydenburg and Gold Mining Estates and Director at Corner House from 1902 with Lionel Phillips, their leading man in Johannesburg. He was a pioneer of hygiene and safety methods in mining and directed much of the Chamber of Mines' administrative policies towards the Chinese.

Appendix: List of Key Figures 193

Sir George Herbert Farrar, 1st Baronet (1859–1915) b. Cambridgeshire. In 1879 he went to South Africa to represent his family's engineering firm. He was sentenced to death after the Jameson Raid, served in the South African War, was knighted in 1902 (the baronetcy was in 1911), and after the war, was Chairman of the East Rand Proprietary Mines (ERPM) from 1893. He became the chief advocate of Chinese labour post-war; sat on the Transvaal Legislative Council, 1903–05; Chairman of the Chamber of Mines in 1904; leader of the Progressive Party in Transvaal, representing Boksburg East, then founded the Unionist Party.

Sir James Percy Fitzpatrick (1862–1931) b. Kingwilliamstown, the Cape. He was sentenced to death for his part in the Jameson Raid, wrote *The Transvaal from Within* (1899) arguing for the need to go to war to protect uitlander rights and later wrote *Jock of the Bushveld* (1907). He was a partner in Eckstein & Co. (1902–07); President, Chamber of Mines (1903), championed white unskilled labour, but supported Chinese importation. In 1906, he retired from mining to devote himself to politics, working as Deputy Leader for the Transvaal Progressives and sitting in the Union Parliament (1911–20) for Pretoria, South Central. He was the first man on the Rand to receive a knighthood in 1902.

Professor Henry Eardley Stephen Fremantle (1874–1931) b. Hatfield, Herts. After getting an MA at Oriel College, Oxford, he became the first Professor of English and Philosophy at the South African College, Cape Town, between 1899 and 1904. He was editor of *South African News* (1903–08, 1910–11). He was a Member of the Legislative Assembly in the Cape (1906–10) for Uitenhage. He was founder and first chairman of the National Party, Cape Province in 1915. He helped organise the South African Party campaign against Chinese labour.

Mohandas (Mahatma) Karamchand Gandhi (1869–1948) b. Gujarat, India. After qualifying in London as a lawyer in 1891, he went to work in Natal in 1893 and began representing the Indian community against the colony's anti-Asiatic legislation. He did the same in the Transvaal. He published *Indian Opinion*, which campaigned against Chinese indentured labour.

John Gardiner Hamilton (1859–?) b. Ireland. After working for the General Post Office, London, he arrived in the Transvaal in 1899. He served as Vice-President of the Chamber of Mines and Chairman and Director of several Transvaal companies. He fought in the South African War and was sent to China to assist Samuel Evans with negotiations and to generally assess the supply of Chinese labourers available. In 1907, he won the Springs seat in the Transvaal as a Progressive, also standing as the Progressive Whip.

William Hosken (1851–1925) b. Hayle, Cornwall. He was a Natal merchant originally (1874–89), then moved to Johannesburg, primarily selling ammunition, dynamite, and arms. He was arrested for his involvement in the Jameson Raid, though fled to Natal during the war. He served as President of the Johannesburg Chamber of Commerce, 1896–1902, then President of the Association of the Chamber of Commerce of South Africa, 1904–05. He also was appointed by Milner to sit in the Transvaal Legislative Council, 1903–07. He sat in the Legislature for Von Brandis as a Progressive from 1907 and was also a member of the Liberal Party in Britain. He supported Gandhi's *Satyagraha* efforts in the Transvaal, and wished for the Transvaal to adopt the Cape's suffrage requirements.

Appendix: List of Key Figures

Henry Charles Hull (1866–1942). A prominent solicitor, he arrived in Kimberley in 1879, before moving to Johannesburg. He was imprisoned for his part in the Jameson Raid, then during the war he was recruited for various cavalry divisions. He was appointed by Milner as an unofficial member of the Legislative Council in the Transvaal, 1903–07. He was one of the four who voted against Farrar's indentured labour importation proposal. He won a seat for Botha's South Africa Party in Georgetown, 1910–12, serving as Minister of Finance.

Sir James Rose Innes (1855–1942) b. Uitenhage, the Cape. He was admitted to the Cape Town bar in 1878, and in 1890 became a QC. He retired from politics in 1902, was knighted in 1901 and acted as a Justice of the High Court in the Transvaal; from union, he became a judge of appeal, and in 1914, became chief justice of South Africa. His attitudes towards Chinese indentured labour were ambivalent.

John Tengo Jabavu (1859–1921) b. Healdtown, the Cape. He edited *Imvo Zabantsundu* (*Native Opinion*) from 1884 in King Williamstown. He campaigned for Merriman, the South African Party, and organised Cape African petitions against Chinese labour.

Sir Leander Starr Jameson (1853–1917) b. Edinburgh. He qualified as a doctor, then went to Kimberley in 1878. He led the Jameson Raid (1895–96) and was sentenced to death. He was leader of the Cape Progressives (1903–10) and Prime Minister of the Cape Parliament, 1904–10. He was made a baronet in 1911.

James William Jamieson (1867–1946) Superintendent of Chinese Labour for the Foreign Labour Department, Transvaal, 1905–08 (seconded). He was also Commercial Attaché to the British Legation in China (1899–1909) and became Consul-General, Canton (1909–26), Tientsin (1926–30) and befriended Sun Yat-Sen. He received a KCMG in 1923.

Hennen Jennings (1854–1920) b. Washington, USA. As an engineer, he worked in Venezuela, then went to Johannesburg and worked for Corner House from 1889. He was first senior mine manager to push for the importation of Chinese indentured labourers.

John Xavier Merriman (1841–1926) b. Somerset, England. He became a member of the Cape Parliament in 1869 and dabbled in journalism and prospecting to raise extra income. He founded the South African Party, which he led, and was Prime Minister (the last) of the Cape (1908–10), continuing in the Union Parliament (1910–24). He campaigned throughout the Cape and used his extensive connections to Liberals throughout the empire to campaign against Chinese importation.

Sir Alfred Milner, 1st Viscount (1854–1925) b. Hesse, Germany. He was a journalist and civil servant, eventually being appointed High Commissioner for South Africa (1897–1905), overseeing the beginning of the South African War. He was Governor of the Transvaal and Orange River Colony (1901–05), leading the official push for Chinese labour importation, though his approval of corporal punishment nearly ruined his career.

William Waldegrave Palmer, 2nd Earl of Selborne (1859–1942). He was Liberal Unionist Chief Whip, Under-Secretary of State for the Colonies (1895–1900), First Lord of the Admiralty (1900–05) and High Commissioner and Governor General for South Africa (1905–16). He was also a friend and admirer of Alfred Milner.

Sir Lionel Phillips (1855–1936) b. London. He went to the Rand in 1889 and worked for Corner House, eventually serving as a senior director, coordinating London and Johannesburg business. He supported British imperialism and was sentenced to death after the Jameson Raid. He was president of the Chamber of Mines in 1892, 1893, 1894, 1895, and 1908. He was made a baronet in 1912.

J. Howard Pim (?–?). He was a leading accountant on the Rand, and was president of the National Democratic Federation, was a close friend of Creswell, campaigned against Chinese labour, and wanted white unskilled labour used instead. He had travelled extensively in the United States and advocated the adoption of the Jim Crow laws in the Southern states.

J. W. Quinn (?–?). A prominent Johannesburg baker and confectioner, he originally opposed to Chinese labour, but eventually voted for it while sitting on the Transvaal Legislative Council (1903–07). He was one of Milner's first appointees to the Johannesburg town council in 1901. He was also Chairman of the African Labour League against any Asian immigration in 1903. He won a seat in the 1907 Transvaal legislature as an independent, largely siding with Het Volk.

Louis Julius Reyersbach (1859–1927). He was Wernher, Beit & Co.'s representative in Kimberley (1894–1900) and partner in Eckstein & Co. from 1902.

Sir Joseph Benjamin Robinson (1840–1929) b. Cradock, the Cape. He made a fortune in Kimberley and on the Rand, but lived mostly in London. He was against the South African War, was friendly with Kruger, and afterwards, politically and financially supported Het Volk. He was made a baronet in 1908, becoming friendly with the Liberal government by being the only mining house to support their cessation of Chinese labour importation.

Raymond Schumacher (?–?). Was an American partner of Eckstein & Company from 1902.

John Alf Sishuba (?–?). He was one of the chief organisers of petitions to the British government against Chinese labour in the Cape.

General The Hon Jan C Smuts, K. C. (1870–1950) b. the Cape. He was a commando general during the South African War. Afterwards, he helped found and write the manifesto for Het Volk and was their Transvaal legislative representative for Wonderboom, and Colonial Secretary (1907–10), overseeing the crushing of the union strikes of 1907 and 1914, as well as Afrikaner revolts in 1914 and 1922. He campaigned extensively against Chinese labour, writing most Het Volk comments on the subject. He became prime minister (1919–24, 1939–48).

Sir Richard Solomon (1850–1913) b. Cape Town. He practised law in the Cape from the 1879, becoming an independent member of parliament for Kimberley throughout most of the 1890s. In 1898, he became Attorney-General, and from 1901, he was legal advisor to the Transvaal government, created a KCMG for his help in drafting the Vereeniging Treaty. From June 1902, he was the Attorney-General to the Transvaal, so drafted much of the legislation relating to Chinese labour. From 1910, he was Agent General for the Transvaal in London.

Sir Julius Wernher (1850–1912) b. Darmstadt, Germany. He arrived in South Africa in 1871, then co-founded Wernher, Beit & Co., directing from London. He occasionally allowed his directors to involve the company in politics.

Peter Whiteside (1870–?) b. Ballarat, Australia. He was an engineer, then engine driver up to 1902. He was in the Transvaal Executive Council of South African Engine Drivers' Association (1897) and General Secretary (1902–19). He was also President of the WTLC (1903–04), sat on the Johannesburg Town

Council (190–196), was the labour union representative on both the 1903 Labour Commission and the Mining Industry Commission, 1907–08. He campaigned against Chinese labour and won a seat as a Labour candidate in 1907.

Wilfred John Wybergh (1868–?) b. Yorkshire. As a mining engineer, he worked as the Commissioner of Mines and member of the Legislative and Executive Councils in Southern Rhodesia from 1891. He later was an engineer to the Consolidated Gold Fields from 1894, before giving up professional work in 1899. He was an Intelligence officer during the war, then was appointed Commissioner of Mines by Milner. He resigned under a cloud, which he always blamed on his opposition to Chinese labour.

Chamber of mines presidents:

1889–92	Hermann Eckstein
1892–96	Lionel Phillips
1896–98	James Hay (resigned in June)
1898–1902	Georges Rouliot
1902–03	Sir Percy Fitzpatrick
1903–04	Sir George Farrar, DSO
1904–05	H. F. Strange
1905–06	F. D. P. Chaplin
1906–07	J. N. de Jongh
1907–08	L. Reyersbach
1908–09	Lionel Phillips
1909–10	J. W. S. Langerman
1910–11	J. G. Hamilton, MVO

Notes

Introduction

1. See in particular D. J. N. Denoon's 'Capitalist Influence and the Transvaal Government During the Crown Colony Period. 1900–1906', *The Historical Journal*, 11:2 (1968), pp.301–331; 'Capital and Capitalists in the Transvaal in the 1890s and 1900s', *Historical Journal*, 23 (March 1980), pp.111–132; *A Grand Illusion: The Failure of Imperial Policy in the Transvaal Colony during the Period of Reconstruction: 1900–1905* (1973); Shula Marks & S. Trapido, 'Lord Milner and the South African State Reconsidered', in M. Twaddle (ed.), *Imperialism, the State and the Third World* (1992); Gary Kynoch, 'Controlling the Coolies: Chinese Mineworkers and the Struggle for Labor in South Africa, 1904–1910', *The International Journal of African Historical Studies*, 36:2 (2003), pp.309–329.
2. George Fredrickson, *White Supremacy: A Comparative Study in American and Southern African History* (1981), p.224.
3. Peter Richardson, *Chinese Mine Labour in the Transvaal* (London, 1982). Also see Coolies and Randlords, 'The North Randfontein Chinese Miners' "Strike" of 1905', *Journal of Southern African Studies*, 2 (1976), pp.151 177; 'The Recruiting of Chinese Indentured Labour for the South African Gold-Mines, 1903–1908', *Journal of African History*, 18:1 (1977), pp.85–107; Richardson, Peter & Jean Jacques Van-Helten, 'Labour in the South African Gold Mining Industry, 1886–1914', in Shula Marks & Richard Rathbone (eds.), *Industrialisation and Social Change in South Africa: African Class Formation, Culture and Consciousness, 1870–1930* (1982), pp.77–98; 'Chinese Indentured Labour in the Transvaal Gold Mining Industry, 1904–1910', in Kay Saunders (ed.), *Indentured Labour in the British Empire, 1834–1920* (1984), pp.260–290.
4. Richardson, 'The Recruiting of Chinese Indentured Labour', p.107.
5. Richardson, 'Recruiting', p.107.
6. Ronald Hyam, *Elgin and Churchill at the Colonial Office, 1905–1908: The Watershed of the Empire-Commonwealth* (1968); J. Marlowe, *Milner-Apostle of Empire: Life of Alfred George Rt. Hon Viscount Milner of St. James' and Cape Town* (1976); D. J. N. Denoon, *A Grand Illusion: The Failure of Imperial Policy in the Transvaal Colony during the Period of Reconstruction: 1900–1905* (1973); G. H. L. Le May, *British Supremacy in South Africa, 1899–1907* (1965); G. B. Pyrah, *Imperial Policy and South Africa: 1902–1910* (1955); A. K. Russell, *Liberal Landslide. The General Election of 1906* (1973); Benjamin Sacks, *South Africa: An Imperial Dilemma-Non-Europeans and the British Nation-1902–1914* (1970).
7. J. Butler & D. Schreuder, 'Liberal Historiography Since 1945', in J. Butler, R. Elphick, & D. Welsh (eds.), *Democratic Liberalism in South Africa: Its History and Prospect* (1987), p.153.
8. Thompson, *Unification*, p.14.

9. Butler & Scheuder, p.153; Thompson, L. M., *The Unification of South Africa, 1902–1910* (1960), p.14.
10. A. Stoler & F. Cooper (eds.), *Tensions of Empire: Colonial Cultures in a Bourgeois World* (1997), pp.vii–viii.
11. Karen Harris, 'Gandhi, the Chinese and Passive Resistance', in J. M. Brown and M. Prozesky (eds.), *South Africa: Principles and Politics* (1996); Harris & F. N. Pieke, 'Integration or Segregation: The Dutch and South African Chinese Compared', in E. Sinn (ed.), *Last Half Century of Chinese Overseas: Comparative Perspectives* (1994). Melanie Yap & Dianne Leong Man, *Colour, Confusion and Concessions: The History of the Chinese in South Africa* (1996).
12. P. C. Campbell, *Chinese Coolie Emigration to Countries within the British Empire* (1923); Alan Jeeves, *Migrant Labour in South Africa's Mining Economy* (1985); D. Northrup, *Indentured Labour in the Age of Imperialism, 1834–1922* (1995); Hugh Tinker, *A New System of Slavery: The Export of Indian Labour Overseas, 1830–1920* (1974).
13. Good examples are J. Evans, P. Grimshaw, D. Philips, & S. Swain, *Equal Subjects, Unequal Rights: Indigenous Peoples in British Settler Colonies, 1830–1910* (2003); Robert A. Huttenback, *Racism and Empire: White Settlers and Coloured Immigrants in the British Self-Governing Colonies, 1830–1910* (1976); Marilyn Lake & Henry Reynolds, *Drawing the Global Colour Line: White Men's Countries and the International Challenge of Racial Equality* (2008).
14. T. R. H. Davenport, *South Africa, A Modern History* (1987), p.226.
15. G. H. Calpin, *There Are No South Africans* (1941), p.9.
16. N. G. Garson, 'English-Speaking South Africa and the British Connection: 1820–1961', in *English-Speaking South Africa Today* (1976), pp.20, 22; Keith Hancock, 'Are There South Africans?', *Hoernle Lecture to the South Africa Institute of Race Relations* (1966).
17. Some of the most notable are Andrew Thompson, 'The Language of Loyalism in Southern Africa, c.1870–1939', *English Historical Review*, 118:477 (June 2003), pp.617–650; Saul Dubow, *A Commonwealth of Knowledge. Science, Sensibility and White South Africa, 1820–2000* (2006), D. E. Omissi & A. S. Thompson (eds.), *The Impact of the South African War* (2002); Hermann Giliomee, *The Afrikaners. Biography of a People* (2003); Jonathan Hyslop, 'The Imperial Working Class Makes Itself "White": White Labourism in Britain, Australia, and South Africa Before the First World War', *Journal of Historical Sociology*, 12:4 (1999), pp.398–421; Peter Merrington, 'Masques, Monuments and Masons: The 1910 Pageant of the Union of South Africa', *Theatre Journal*, 49:1 (1997), pp.1–14; Nill Nasson, 'Why They Fought: Black Cape Colonists and Imperial Wars, 1899–1918', *The International Journal of African Historical Studies*, 37:1 (2004), pp.55–70; A. Odendaal, *Vukani Bantu! The Beginnings of Black Protest Politics in South Africa to 1912* (1984); S. Marks & S. Trapido (eds.) *The Politics of Race, Class and Nationalism in Twentieth Century South Africa* (1987); Paul Rich, *State Power and Black Politics in South Africa, 1912–51*(1996); Peter Walshe, *The Rise of Nationalism in South Africa: The African National Congress, 1912–1952* (1970); L. Thompson, *The Political Mythology of Apartheid* (1985).

18. Here, I use 'nation' and 'nationhood' to describe political identity and 'state' to refer to the legal entity of South Africa. See Frederick Cooper's excellent analysis of the terminology, *Colonialism in Question, Theory, Knowledge, History* (2005), especially Chapter 3; see also Philip L. White, 'Globalisation and the Mythology of the "Nation State"', in A. G. Hopkins, (ed.), *Global History, Interactions between the Universal and the Local* (2006), p.258. For an example of how colonial South African history can be affected, see B. M. Magubane, *The Making of a Racist State: British Imperialism and the Union of South Africa, 1875–1910* (1996).
19. J. G. A. Pocock, 'British History: A Plea for a New Subject', *Journal of Modern History* (175); Carl Bridge & Kent Fedorowich, 'Mapping the British World', *Journal of Imperial and Commonwealth History*, 31:2 (2003), pp.1–15; John Darwin, 'A Third British Empire? The Dominion Idea in Imperial Politics', in Judith M. Brown & Wm. Roger Louis (eds.), *The Oxford History of the British Empire, volume IV: The Twentieth Century* (2001), pp.64–87.
20. Darwin, 'A Third British Empire?', pp.64–87; Bridge & Fedorowich, pp.1–15. On emerging nationalism, see also Douglas Cole, 'The Problem of "Nationalism" and "Imperialism" in British Settlement Colonies', *The Journal of British Studies*, 10:2 (May 1971), pp.160–182; John Eddy & Deryck Schreuder (eds.), *The Rise of Colonial Nationalism, Australia, New Zealand, Canada and South Africa First Assert Their Nationalities, 1880–1914* (1988), pp.19–93, 192–226.
21. Matthew Guterl & Christine Skwiot, 'Atlantic and Pacific Crossings: Race, Empire, and "the Labor Problem" in the Late Nineteenth Century', *Radical History Review*, 91 (Winter 2005), p.42.
22. Stoler & Cooper, *Tensions of Empire: Colonial Cultures in a Bourgeois World* (1997), p.28. For instance, see Guterl & Skwiot; Lake & Reynolds.
23. Tony Ballantyne, 'Race and the webs of empire', p.2.
24. For recent, but still rare attempts, see Tony Ballantyne, *Orientalism, Race, Aryanism in the British Empire* (2002); Z. Laidlaw, *Colonial Connections, 1815–45: Patronage, the Information Revolution and Colonial Government* (2005); Alan Lester & David Lambert (eds.), *Colonial Lives across the British Empire: Imperial Careering in the Long Nineteenth Century* (2006).
25. Deryck M. Schreuder & Stuart Ward (eds.), *Australia's Empire* (2008); P. Buckner (ed.), *Canada and the British Empire* (2008).
26. Ballantyne, 'Race', p.38.
27. A. S. Thompson, 'The Languages of Loyalism in Southern Africa', *English Historical Review*, 117: 477 (June 2003), p.618.
28. F. Cooper, *Colonialism in Question: Theory, Knowledge, History* (2005), p.18.
29. The most eloquent critique of the term 'non-white' can undoubtedly be found in the writings and speeches of Steve Biko, See *I Write What I Like* (1978).
30. Stephen Ellis, 'Violence and History: A Response to Thandika Mkandawire', *The Journal of Modern African Studies*, 41:3 (September, 2003), p.469.

200 Notes

1 Chinese Migration and 'White' Networks, c.1850–1902

1. Bridge & Fedorowich, p.4.
2. See, for instance, Lester, *Imperial Networks*.
3. The term 'Asian' will be used throughout, because it was commonly used within the empire, Europe, and the United States to refer to Indians, Chinese, and later, Japanese as the three largest pools of migrants; it could on occasion include Afghans, Micronesians, and Polynesians.
4. See, for instance, Thomas C. Holt, *The Problem of Freedom: Race, Labor and Politics in Jamaica and Britain, 1832–1938* (1992), p.236.
5. Patrick Manning, *Migration in World History* (New York, 2005), p.149.
6. A. McKeown, 'Global Migration 1846–1940', *Journal of World History*, 19:2 (2004), p.158; *Melancholy Order: Asian Migration and the Globalization of Borders* (2008).
7. The derogatory term 'coolie' was usually applied to indentured or unskilled workers on long-term restricted contracts for work on plantations and probably derived from the Tamil word *kuli*, meaning 'wages'.
8. G. V. Doxey, *The Industrial Colour Bar in South Africa* (1961), p.8.
9. Northrup, p.22. Also see Walton Look Lai, *Indentured Labor, Caribbean Sugar: Chinese and Indian Migrants to the British West Indies, 1838–1918* (1993).
10. Yen Ching-Hwang, *Coolies and Mandarins* (1985), p.46. For further information about who migrated from Asia, see Pieter Emmer, 'Freedmen in the Caribbean, 1834–1917', in Howard Temperley (ed.), *Emancipation and its Discontents* (2000), p.159; McKeown, 'Global', p.171.
11. P. Richardson, 'The Natal Sugar Industry in the Nineteenth Century', in W. Beinart, P. Delius, & S. Trapido (eds.), *Putting a Plough to the Ground: Accumulation and Dispossession in Rural South Africa* (1986), pp.129–175; Maureen Tayal, 'Indian Indentured Labour in Natal, 1890–1911', *The Indian Economic and Social History Review*, 14:4 (1977), pp.519–547.
12. A breakdown of the historiography can be found in A. McKeown, *Melancholy Order: Asian Migration and the Globalization of Borders* (2008), pp.66–89. See also 'Conceptualizing Chinese Diasporas, 1842 to 1949', *The Journal of Asian Studies*, 58:2 (May 1999), esp. pp.312–316; Wang Gungwu, *China and the Chinese Overseas* (1991), 'Introduction'.
13. McKeown, 'Global', p. 175.
14. Huguette Li-Tio-Fane Pineo, *Chinese Diaspora in Western Indian Ocean* (1985), translation from French, p.2.
15. Richardson, 'The Recruiting of Chinese Indentured Labour', p.107.
16. For more on the general theory of stereotype formation, see Sander S. Gilman, *Difference and Pathology, Stereotypes of Sexuality Stereotypes of Sexuality, Race and Madness* (1985), especially pp.15–16.
17. Stuart Creighton Miller, *The Unwelcome Immigrant: The American Image of the Chinese, 1785–1882* (1969).
18. Colin Mackerras, *Sinophiles and Sinophobes: Western Views of China* (2000), p.60.
19. Karen L Harris, 'Sugar', pp.147–158; Brian Kennedy, *A Tale of Two Cities. Johannesburg and Broken Hill: 1885–1925* (1984); Hyslop, 'Imperial', pp.398–421; John Higginson, 'Privileging the Machines: American

Engineers, Indentured Chinese and White Workers in South Africa's Deep-Level Gold Mines, 1902–1907', *IRSH* 52 (2007); Guterl & Skwiot.
20. Simon J. Potter, *News and the British World: The Emergence of an Imperial Press System, 1876–1922* (2003), p.20.
21. Potter, p.14.
22. While not all stories were carried in this way, stories about the Chinese were particularly applicable to multiple spaces. For more on the importance of Reuters in shaping information networks through news replication about the South African scheme across the English-speaking world, see Chapters 3 and 5. See also the Reuters Archives in London; Donald Read, 'Reuters and South Africa: "South Africa is a country of monopolies"', *South African Journal of Economic History*, 11:2 (1996), pp.104–143.
23. Gilman, p.20.
24. Governor Sir Charles Fitzroy, to Grey, 3 October 1849, quoted in Huttenback, *Racism*, p.29.
25. See David Gouter, *Guarding the Gates: The Canadian Labour Movement and Immigration, 1872–1934* (2007).
26. Campbell, p.34.
27. Andrew Markus, *Fear and Hatred: Purifying Australia and California, 1850–1901* (1979), p.10.
28. Hyslop, 'Imperial', pp.405–406. See also Gouter; Andrew Thompson, *The Empire Strikes Back? The Impact of Imperialism on Britain from the Mid-Nineteenth Century* (2005), p.69.
29. *Cape Argus*, 24 February 1876.
30. Hyslop, 'Imperial', pp.405–406.
31. Many of these settlers were also against European migrants who might compete. See Avner Offer, *The First World War: An Agrarian Interpretation* (1989), p.179.
32. Markus, p.1.
33. Markus, pp.14–15, 24, 33.
34. Guterl & Skwiot, p.43.
35. Alexander Saxton, *The Indispensible Enemy, Labor and the Anti-Chinese Movement in California* (1975), p.62.
36. June 1869, miner's union of Virginia City & Golden Hill in address to working men of Nevada, quoted in Saxton, p.59. See also Markus, pp.10, 12.
37. Markus, p.xviii.
38. G. B. Densmore, *The Chinese in California* (1880), pp.117–119.
39. Some historians have also emphasised that racialised notions of suffrage were based on earlier US debates about republicanism. See David R. Roediger, *The Wages of Whiteness: Race and the Making of the American Working Class* (1991); Erin L. Murphy, ' "Prelude to imperialism": Whiteness and Chinese Exclusion in the Reimagining of the United States', *Journal of Historical Sociology*, 18:4, (2005), p.460.
40. Lake & Reynolds, p.6. See also Lake's 'The White Man under Siege: New Histories of Race in the Nineteenth Century and the Advent of White Australia', *History Workshop Journal*, 58, (1994), p.53.
41. Kennedy, p.44; Campbell, p.64; Markus, p.21.
42. Quoted in Kennedy, p.44.

43. Jules Becker, *The Course of Exclusion, 1882–1924: San Francisco Newspaper Coverage of the Chinese and Japanese in the United States* (1991).
44. Jane Samson (ed.), *Race and Empire* (2005), p.73. Some Chinese even welcomed such fear. See Gregory Blue, 'Gobineau on China: Race Theory, the "Yellow Peril", and the Critique of Modernity', *Journal of World History*, 10:1 (1999), pp.129–130.
45. See David Walker, 'Australia's Asian Futures', in Martyn Lyons & Penny Russell (eds.), *Australia's History: Themes and Debates* (2005); Markus; Lake, 'Translating Needs into Rights: The Discursive Imperative of the Australian White Man, 1901–30', in Stefan Dudink, Karen Hagemann, & John Tosh (eds.), *Masculinities in Politics and War: Gendering Modern History* (2004), p.206 and Lake, 'The White Man under Siege', 41; Offer, *The First World War*, p.210.
46. Campbell, pp.80–82.
47. Huttenback, *Racism*, pp. 94–95; Markus, pp.227–228.
48. Huttenback, p.31.
49. Quoted in Campbell, p.65.
50. Quoted in Gorman, 'Wider and Wider Still?', 13. Italics are Gorman's.
51. Quoted in Markus, p.xv.
52. Walker, 'Australia's', pp.98–126.
53. See, for instance, John Seed, 'Limehouse Blues: Looking for Chinatown in the London Docks, 1900–40', *History Workshop Journal*, 62 (2006), p.64; Gregor Benton & Edmund Terence Gomez, *The Chinese in Britain, 1800-Present, Economy, Transnationalism, Identity*(2008); Shompa Lahiri: *Indians in Britain: Anglo-Indian Encounters, Race and Identity, 1880–1930* (2000).
54. Quoted in Christine Bolt, *Victorian Attitudes to Race* (1971), p.200.
55. Bolt, p.144.
56. A. T. Yarwood, *Asian Migration to Australia: The Background to Exclusion 1896–1923* (1964), p.11; Markus, pp.184–185.
57. Ballantyne, *Orientalism*, pp.80–81.
58. Quoted in Jeremy Martens, 'A Transnational History of Immigration Restriction: Natal and New South Wales, 1896–97', *Journal of Imperial and Commonwealth History*, 34:3 (2006), p.322.
59. Martens, 'transnational', p.334.
60. See ch.1 in Eddy & Schreuder.
61. Robert W. D. Boyce (February 2000), 'Imperial Dreams and National Realities: Britain, Canada and the Struggle for a Pacific Telegraph Cable, 1879–1902', *The English Historical Review*, 115:460, pp.39–70.
62. UKPP Cd8596, Proceedings of a Conference between the Secretary of State for the Colonies and the Premiers of the Self-Governing Colonies, at the CO, London, June and July 1897, pp.13–14.
63. Robert Huttenback, 'The British Empire as a "White Man's Country" – Racial Attitudes and Immigration Legislation in the Colonies of White Settlement', *Journal of British Studies*, 13 (1973), p.111.
64. Huttenback, 'British', p.108.
65. Ballantyne, *Orientalism*, pp.80–81.
66. See for examples: Robert A. Huttenback, *Gandhi in South Africa, British Imperialism and the Indian Question, 1860–1914* (1971); Harris, 'Gandhi, the Chinese and Passive Resistance'; Andrea Geiger, 'Negotiating the Boundaries of Race

and Class: Meiji Diplomatic Responses to North American Categories of Exclusion', *B.C. Studies*, 156/157 (Winter/Spring 2007).
67. Offer, p.196.
68. Marilyn Lake, 'Translating', p.203.
69. Campbell, p.78.
70. For a wider discussion of this, see Eddy & Schreuder.

2 The Transvaal Labour 'Problem' and the Chinese Solution

1. Charles Sydney Goldman, 'South Africa and Her Labour Problem', *Nineteenth Century & After*, 4 (May 1904), p.848.
2. See Ben Mountford, 'Australia's Empire and the Chinese Question', unpublished paper, March 2009; Kwabena O. Akurang-Parry, ' "We Cast about for a Remedy": Chinese Labor and African Opposition in the Gold Coast, 1874–1914', *The International Journal of African Historical Studies*, 34:2 (2001), pp.365–384.
3. See, for instance, Millicent Garrett Fawcett, 'Impressions of South Africa, 1901 and 1903', *Contemporary Review*, 84 (July/December 1903), p.647.
4. R. Ross, *Beyond the Pale: Essays on the History of Colonial South Africa* (1993), p.48.
5. See UKPP Cd1897, Reports of the Transvaal Labour Commission, Minutes of Proceedings and Evidence(1904). Such a view was, of course, common throughout European colonisation of Africa. See Cooper, Holt, & Scott, *Beyond Slavery*; Crush, Jonathan, Alan Jeeves, & David Yudelman, *South Africa's Labor Empire: A History of Black Migrancy to the Gold Mines* (1991).
6. J. T. Darragh, 'The Native Problem in South Africa', *Contemporary Review*, 81 (January/June 1902), p.97. See also Sir Alfred Pease, late Administrator of Native affairs in the Transvaal, 'The Native Question in the Transvaal', *Contemporary Review*, 90 (July/December 1906), pp.16–36; Jerome E. Dyer, 'The South African Labour Question', *Contemporary Review*, 83 (January/June 1903), pp.439–445; Alfred F. Fox, et al., 'The Native Labour Question in South Africa', *Contemporary Review*, 83 (January/June 1903), pp.540–553; *Rt Hon Sir William Harcourt, MP, 'Native Labour in South Africa. Two Letters to the Times', Reprinted by the Liberal Publication Dept. (1903)*: Letter II. 16th February 1903.
7. Denoon, 'Transvaal Labour', p.487.
8. David Kynaston, *The City of London, volume II, Golden Years, 1890–1914* (1995), pp.105–132.
9. CO291/68/4234, Report of the Transvaal Treasury on the financial year 1902–03 by the Transvaal Chief Inspector of Revenue to the Colonial Treasurer, 9 January 1904.
10. See Crush, Jeeves, & Yudelman, pp.4–8.
11. P. Harries, *Work, Culture and Identity: Migrant Labourers in Mozambique and South Africa, 1860–1910* (1994).
12. Richardson, *Chinese*, p.11.
13. Diana Cammack, *The Rand at War, 1899–1902, The Witwatersrand & the Anglo-Boer War* (1990).
14. Stanley Trapido, 'Landlord and Tenant in a Colonial Economy: The Transvaal 1880–1910', *Journal of Southern African Studies*, 5 (October 1978), pp.26–58;

Charles Van Onselen, 'Race and Class in the South African Countryside: Cultural Osmosis and Social Relations in the Sharecropping Economy of the South Western Transvaal, 1900–1950', *The American Historical Review*, 95:1 (February 1990), pp.99–123.
15. Richardson, *Chinese*, pp.18–20. See UKPP Cd1895 and Cd1896 for their statistics and a list of past recruiting efforts. See also HE251/138/395, Report on the Wastage of Native Labour in the Mines of the Witwatersrand District by T. J. Britten, 1903.
16. CO291/47/36082, Milner to Chamberlain, 28 August 1902; FO to CO, 29 August 1902.
17. The Consolidated Gold Fields, p.77.
18. W. Nimocks, *Milner's Young Men: The 'Kindergarten' In Edwardian Imperial Affairs* (1968).
19. Cell, pp.192–193.
20. CO417/326/42283, Milner to Chamberlain, 8 November 1901, quoted in Denoon, 'Transvaal Labour', p.481. See also Cd1552, Milner to Chamberlain, 8 September 1902, p.8; Van-Helton, J. J. & Keith Williams, ' "The Crying Need of South Africa": The Emigration of Single British Women to the Transvaal, 1901–10', *Journal of Southern African Studies*, 10 (October 1983), pp.17–38.
21. Milner to Hanbury Williams, CO, February 1901, quoted in Headlam, p.217.
22. Richard D. Dawe, *Cornish Pioneers in South Africa: Gold and Diamonds, Copper and Blood* (1998), pp.193–224.
23. Milner Papers (MP)2/217, Milner to Hanbury Williams, CO, February 1901. See also Ian R. Smith, 'Capitalism and the War', in Omissi & Thompson, pp.56–75.
24. Giliomee, *Afrikaners*, pp.236–239.
25. *Cape Times*, 29, 30 December 1903; see also The London *Times*, To the Editor, from Carl Meyer, 16 March 1906, p.5; Extract, A. Bailey to John X. Merriman, 11 November 1903, in P. Lewsen (ed.), *Selections from the Correspondence of John X. Merriman, volume III, 1905–1924* (1969), p.405.
26. For a more detailed analysis of this work and links between the government and mines, see Jeeves, *Migrant Labour*, pp.48–49; Denoon, 'Capitalist Influence', pp.301–331.
27. CO205/345 and UKPP Cd2025, Correspondence Relating to the Conditions of Native Labour Employed in the Transvaal Mines, 1904.
28. UKPP Cd2104, Correspondence (containing the Chinese Labour Ordinance) relating to Affairs in the Transvaal and Orange River Colony, Lawley to Milner, 14 May 1904, pp.6–7.
29. UKPP Cd1896, Transvaal Labour Commission Majority Report, p.33.
30. UKPP Cd1894, p.31. Testimony of Mr. Mellow Breyner.
31. Headlam, Lord Milner to Mr. Chamberlain, 6 April 1903, p.461.
32. *Rand Daily Mail*, 27 February 1903; See also Warwick, *Black*, p.17; Long, *Drummond Chaplin* (London, 1941), p.57; Alfred F. Fox, et al., 'The Native Labour Question in South Africa', *Contemporary Review*, 83 (1903:January/June), pp.542–543.
33. Report, Commissioner of Native Affairs, 1902.
34. Keegan, *Rural*, pp.23–24.

35. Edgar P. Rathbone, 'The Native Labour Question', *Nineteenth Century and After,* September (1903), pp.406–407; HE202/12, Weirner, Beit & Co. to Eckstein & Co., 27 March 1903.
36. Richardson, *Chinese,* p.16. See the 1903 Labour Commission Report and evidence, UKPP Cd1894–7; HE71, Wernher, Beit & Co. to Eckstein & Co., 10 October 1902.
37. Jeeves, p.70.
38. Evidence of Silas Crowle, 1 July 1907, Mining Industry Commission, 1907–08, pp.310–311, referenced in Kennedy, p.25.
39. Trapido, 'Landlord', 51.
40. Wheatcroft, *Randlords,* p.219; see also Paul H. Emden, *Randlords* (1935), p.175.
41. For his views on white labour, see F. H. P. Creswell, *The Chinese Labour Question from Within – Facts, Criticisms, & Suggestions – Impeachment of a Disastrous Policy* (1905), p.82; Kennedy, pp.23–24; Katz, *Labour,* pp.147–149. He had no known links to labour unions at this time.
42. HE251/138/361, W. Beit & Co. to H. E. & Co., 26 November 1903; TAR/Half Year Report of TGME, December 1905.
43. Ticktin, p.68.
44. UKPP Cd1896, Reports of the Transvaal Labour Commission, Minority Report, Appendix, Mr. Terbutt's Letter, 3 July 1902, p.65; F. D. P. Chaplin, 'The Labour Question in the Transvaal', *National Review* (April 1903), pp.296–297; Charles Rudd, *The Times,* letter, 16 February 1903; *SAN,* 20 August 1904.
45. Katz, 'Underground', p.476. Also see Davies, pp.41–69.
46. Report of Executive Committee of Transvaal Chamber of Mines, Chamber of Mines *Annual Report,* 1903, p.30, my italics. See also evidence at the 1903 Labour Commission, UKPP Cd1897, pp.608–609; Chamber of Mines *Annual Report* 1903, p.L, pp.153–154; R. E. Browne, 'Working Costs of the Mines of the Witwatersrand', *Journal of the South African Institution of Engineers,* 12 (1907), pp.297–298.
47. Phillips, *Labour,* p.64.
48. Johannesburg *Star,* 20 June 1902; *Rand Daily Mail,* 27 February 1903, Chamber of Mines Annual Meeting 1903, Sir J. P. Fitzpatrick's Speech.
49. UKPP Cd1895, 2 June 1903, pp.37–44.
50. L. S. Amery, *My Political Life, volume I* (1953), p.175. See also Belinda Bozzoli, *The Political Nature of a Ruling Class: Capital and Ideology in South Africa, 1890–1933* (1981), pp.90–93. For a sympathetic analysis of his life written by his daughter, see Margaret Creswell, *An Epoch of the Political History of South-Africa in the Life of Frederic Hugh Page Creswell* (1956), pp.33–51.
51. G. Burke & P. Richardson, 'The Profits of Death: A Comparative Study of Miners' Phthisis in Cornwall and the Transvaal, 1876–1918', *Journal of South African Studies,* 4:2 (1978), p.151.
52. Chamber of Mines *Annual Report* 1898, pp.4–5.
53. *Bloemfontein Post,* 6 January 1903. See also Monypenny to Moberly Bell, 20 March 1899, quoted in Porter, 'Milner', p.335; V. G. Kiernan, *The Lords of Human Kind: European Attitudes Towards the Outside World in the Imperial Age* (1969), p.225.

54. Vivian Bickford-Smith, *Ethnic Pride and Racial Prejudice in Victorian Cape Town, Group Identities and Social Practice, 1875–1902* (1995), p.148.
55. H. O'Kelly Webber, *The Grip of Gold, The Life Story of a Dominion* (1936), p.131.
56. See HE71, W & B to E & Co., Private, 13 March 1903; HE202/12, Labour (Private), Weirner, Beit & Co. to Eckstein & Co., 27 March 1903; HE208/12, Wernher, Beit & Co. to Eckstein & Co., 3 April 1903; Chaplin, 'The Labour', p.297; Mawby, 'Capital', p.394; Mawby, *Gold II*, pp.363–364
57. *Rand Daily Mail*, 27 February 1903; CO291/53, Robert Fricker to Lord Harris, 3 November 1902; A. H. Duminy & W. R. Guest (eds.), *Fitzpatrick, South African Politician, Selected Papers, 1888–1906* (1976), Fitzpatrick to J. Wernher, 25 July 1902, pp.340–341; Mawby, 'Capital', p.394; Long, p.59; Grant, p.85.
58. HE202/12, Weirner, Beit & Co. to Eckstein & Co., 27 March 1903. The text in italics was originally underlined in the letter.
59. For more about Farrar's mining interests, see Wheatcroft, *Randlords*, pp.158–159; R. V. Kubicek, *Economic Imperialism in Theory and Practice: The Case of South African Gold Mining Finance, 1886–1914* (1979), p.133; Mawby, *Gold Mining vol. I*, pp.359–360; Farrar Papers (FP), Rhodes House.
60. Sascha Auerbach, *Race, Law, and 'The Chinese Puzzle' in Imperial Britain* (2009), p.28.
61. Headlam, Milner to Just, 14 June 1903, pp.463–464. The financial figures are in UKPP Cd1552 and Cd1586.
62. Headlam, Milner to Chamberlain, 6 April 1903, p.461; see also Milner to Dr. J. E. Moffat, 1 April 1903, p.460.
63. Headlam, Milner to E. H. Walton, 8 April 1903, p.461.
64. John Buchan, *Memory Hold the Door* (1940), p.100; see a similar assessment in L. S. Amery, *My Political Life, volume I* (1953), p.174.
65. UKPP Cd1640, Minutes of Proceedings of the South African Customs Union Conference held at Bloemfontein, March 1903, Thursday, 19 March 1903, p.13; see also UKPP Cd1599.
66. See CO417/381; Ian Phimister, An Economic and Social History of Zimbabwe, 1890–1948: Capital Accumulation and Class Struggle (1988), pp.45–49 for more information on Rhodesia's labour situation.
67. HE208/12, Reyersbach to Wernher, 2 August 1902; Cd1895, Enclosure in No.87, p.149; HE71, General Correspondence, 13 March 1903, W & B to E & Co., Private.
68. UKPP Cd1683, Milner to Chamberlain, 12 May 1903, pp.3–4; reply, 23 May 1903, pp.4–5; Headlam, p.438.
69. *Rand Daily Mail*, 6 June 1903.
70. UKPP Cd1683, Chamberlain to Milner, 23 May 1903, pp.4–5.
71. UKPP Cd2028, CO to Rhodesian Land and Mine Owners' Association, 2 December 1903, p.8; Mutanbirwa, pp.101–102.
72. UKPP Cd1897, Testimony to the Transvaal Labour Commission, p.2. Copy of advertisement.
73. UKPP Cd1894, Majority Report of the Labour Commission, p.9; *SAN*, 20 July 1903.
74. UKPP Cd1896, Majority Report, p.33.

75. See UKPP Cd1897, Reports of the Transvaal Labour Commission, Minutes of Proceedings and Evidence,1904 for the various testimonies about why African labour was proving so elusive.
76. See UKPP Cd1897, Testimony of General Louis Botha, p.274. Some British colonials also promoted this idea; see Rand Daily Mail, 6 June 1903.
77. Cd1894 Minority Report, p.53.
78. Rand Daily Mail, 3 June 1903.
79. UKPP Cd1894, Minority Report, p.57.
80. UKPP Cd1895, Chamber of Mines meeting, 2 December 1903, p.156.
81. HE250/140, Skinner to WNLA, 13 June 1903.
82. S. Evans to F. Eckstein, 22 June 1903, quoted in Mawby, 'Capital', p.395.
83. HE208/12, unknown to Eckstein, 28 September 1903.
84. UKPP Cd1895, Report of H. Ross Skinner, 22 September 1903, p.86. See also Chamber of Mines *Annual Report* (1903).
85. UKPP Cd1895, Skinner Report, pp.77–78.
86. Chamber of Mines Annual Report (1903), pp.xxx–xxxi, p.169.
87. UKPP Cd1895, Skinner Report, p.82; See Chapters 4 and 6 for more on segregation.
88. UKPP Cd1895, De Jongh statement, Chamber of Mines meeting, 2 December 1903, p.155; *Star*, 28 December 1903, p.265.
89. UKPP Cd2026, Milner to Lyttelton, 13 April 1904, p.37.
90. UKPP Cd2104, Lawley to Milner, 14 May 1904, p.6.
91. See UKPP Cd1956, Ordinance No. 17(1904). For more on the controversy, see Chapters 3 and 4.

3 Greater Britain in South Africa: Colonial Nationalisms and Imperial Networks

1. Headlam, Minute, 20 January 1903, p.438. See also Milner to Chamberlain, 6 April 1903, p.461; UKPP Cd2104, Lyttelton to Australian Governor-General, Lord Northcote, 20 May 1904, p.3; LtG111/101/20, Lawley to Harger, 10 February 1903.
2. Johannesburg *Star*, 1 April 1903; George Farrar, 'The South African Labour Problem. Speech by Sir George Farrar at a Meeting Held on the East Rand Proprietary Mines, On 31 March 1903' (1903).
3. Van der Horst, pp.171–172.
4. Bala Pillay, *British Indians in the Transvaal. Trade, Politics, and Imperial Relations, 1865–1906* (1976), p.17.
5. *Transvaal Leader*, 2 April 1903. For more on the White League, see *Indian Opinion*, 23 July 1903; Pillay, p.152.
6. *Transvaal Leader*, J. Quinn speech, 2 April 1903.
7. M. C. Bruce, *The New Transvaal* (1908), p.90.
8. Saul Dubow, 'Colonialism, Imperialism, Constitutionalism', in *A Commonwealth of Knowledge. Science, Sensibility and White South Africa, 1820–2000* (2006), pp.121–157; Saul Dubow, 'Imagining the New South Africa in the Era of Reconstruction', in Omissi & Thompson (eds.), pp.56–75.
9. For background, see Timothy Keegan, *Colonial South Africa and the Origins of the Racial Order* (1996); Ross, *Beyond*.

10. See, for instance, *Indian Opinion*, 21 January, 21 May 1904; Les Switzer, 'Gandhi in South Africa: The Ambiguities of Satyagraha', *Journal of Ethnic Studies*, 14:1 (1986), p.125 for debates about legal definitions of 'coloured' and 'native'.
11. See J. Eddy, D. Schreuder, & Simon Potter, 'Richard Jebb, John S. Ewart and the Round Table, 1898–1926', *English Historical Review*, CXXII: 495 (2007), pp.105–132; Daniel Gorman, 'Lionel Curtis, Imperial Citizenship, and the Quest for Unity', *The Historian*, 66:1 March 2004, pp.67–96; J. Darwin, 'A Third British Empire?'; J. Kendle, *The Round Table Movement and Imperial Union* (1975); A. Bosco & A. May (eds.), *The Round Table: The Empire/Commonwealth and British Foreign Policy* (1997); N. Mansergh, *The Commonwealth Experience, Volume I, the Durham Report to the Anglo-Irish Treaty* (1982 [1969]); Max Beloff, *Imperial Sunset, volume I, Britain's Liberal Empire, 1897–1921* (1969), pp.66–84; Ged Martin, 'The Idea of "Imperial Federation"', in Ronald Hyam & Ged Martin, *Reappraisals in British Imperial History* (1975), pp.121–138.
12. Omissi and Thompson; Lowry and R. J. D. Page, 'Canada and the Imperial Idea in the Boer War Years', *Journal of Canadian Studies*, 1:1 (February 1970), pp.33–49; Carman Miller, *Painting the Map Red: Canada and the South African War, 1899–1902* (1993); Craig Wilcox, *Australia's Boer War: The War in South Africa 1899–1902* (2002); Carl Bridge & Bernard Attard (eds.), *Between Empire and Nation: Australia's External Relations from Federation to the Second World War* (2000); R. A. Crouch, 'An Australian View of the War', *Contemporary Review*, 86 (July/December, 1904), p.181.
13. See Rachel Bright, 'Asian Migration and the British World, 1850–1914', in Andrew S. Thompson and Kent Fedorowich (eds.), *Empire, Identity and Migration in the British World* (2013).
14. Herbert Samuel, 'The Chinese Labour Question, *Contemporary Review* 85 (January/June 1904), pp.457–467.
15. A. N. Porter, 'Milner', p.324.
16. Kenneth O. Morgan, 'The Boer War and the Media (1899–1902), *Twentieth Century British History*, 13:1 (2002), 2; Porter, 'Milner', p.339.
17. Andrew Thompson, 'Imperial Propaganda during the South African War', in Cuthbertson et al. (eds.), *Writing a Wider War* (2002), p.304. See also Arthur Davey, *The British Pro-Boers, 1877–1902* (1978), pp.71–72.
18. Thompson, 'Imperial Propaganda', p.315.
19. Thompson, 'Imperial Propaganda', pp.318–319. See Annual Report of the ISAA (1904–05), (1905–06).
20. Mr John Stroyan, MP, 'A Scottish South African MP on the Importation of Chinese Labourers', ISAA #66.
21. A Jabez Strong, 'The Transvaal Labour Question. A British Workingman to British Workingmen. What it all Means to British Labourers', ISAA #70.
22. Anonymous, 'Chinese Labour: Five Reasons for Supporting the Government on Chinese Labour', ISAA #60; see also MP341/12, ISAA pamphlet: 'Asiatic Labour in the Transvaal', n.d.
23. LtG102/101/25, CMLIA to Lieutenant-Governor, 13 June 1903; UKPP Cd1941, James B. Macdonald, Secretary, CMLIA, to Lieutenant-Governor, 23 September 1903, p.26 (signed by 43 'residents').

24. 'Notes on the Labour Position in the Transvaal by the London Secretary of the Transvaal Chamber of Mines' (London, February 1904).
25. See *The Times*, 26 August 1903; 25 February 1904; 1,9 April 1904, 24 May 1904. Those against also wrote letters; see Botha, 15 July 1903 (also in *de Volksstem*, 22 August 1903); letter from John Seely, 12 January 1904; from Creswell, 1 February 1904.
26. Chaplin, pp.296–297; Charles Rudd, *The Times*, letter, 16 February 1903.This was done again in 1905–06, with Creswell and Schumacher.
27. Creswell, *Chinese*, p.55; Pyrah, p.193.
28. Rykie Van Reenen (ed.), *Emily Hobhouse: Boer War Letters* (Pretoria, 1984), Emily Hobhouse to Lady Hobhouse, 17 September 1903, p.290, my italics.
29. British Colonist, 'British Rule in the Transvaal, *Contemporary Review*, 85 (January/June 1904), p.348.
30. *The Sydney Morning Herald*, 26 March 1904, p.7.
31. *Rand Daily Mail*, 18 January 1904.
32. *Transvaal Leader*, 5 December 1903; UKPP Cd1895, Lawley, Lieutenant-Governor, to Lyttelton, 12 December 1903, p.179.
33. UKPP Cd1941, Milner to Lyttelton, 15 February 1904, p.23; Chamber of Mines to Lieutenant-Governor, 19 January 1904, p.13.
34. UKPP Cd1895, Johannesburg Stock Exchange to Lieutenant-Governor, Transvaal, 10 December 1903, p.179.
35. *Rand Daily Mail*, 18 January 1904.
36. ISAA, Chinese Labour for the Transvaal Mines: Nonconformists Condemn Agitation Against It (London, 1904). See also Conservative Publication Department, *Chinese Labour, A Protest from Transvaal Nonconformist Ministers* (London, 1904).
37. Kevin Grant, *A Civilised Savagery: Britain and the New Slaveries in Africa, 1884–1926* (2005), pp.79–108; Rev. T. W. Pearce, 'Chinese Coolie Labour in South Africa, A Statement from the Chinese Missionary Standpoint', *Chronicle of the London Missionary Society* (1904), pp.226–228; Rev. Arnold Foster, 'Chinese Coolie Labour in South Africa' (1905), pp.19–23.
38. Chinese labour: dignified rebuke from Nonconformist ministers in the Transvaal to their brethren in England, ISAA, #73, taken from the *Methodist Recorder*, 7 April 1903. The signatories were Amos Burnet, Wesleyan Chairman of the District and Vice-President Church Council, John C. Harris, Congregational Minister and Sect. Church Council, N. A. Ross, Presbyterian Minister and Member of the Executive of Church Council; all of them lived and worked in South Africa; Amos Burnet was a well-known Lincolnshire man and Chairman of Wesleyans Methodist Church in the Transvaal.
39. *Amalgamated Engineering Journal (AEJ)*, especially 2 April 1903 and 'Monthly Prefaces' throughout December 1903, March, April 1904; See also Katz, *Trade*, pp.110–111; Thompson, *Empire*, pp.70–73; *The Advertiser* (Adelaide), 3 February 1904.
40. *Transvaal Leader*, 2 April 1903; Katz, *Trade*, pp. 110–111.
41. *Star*, 2 April 1903.
42. Jacqueline Beaumont, 'An Irish Perspective on Empire: William Flavelle Monypenny', in Simon J Potter (ed.), *Newspapers and Empire in Ireland and Britain* (2004), pp.177–194.

43. *SAN*, 11 July 1903.
44. *Star*, 22 January 1903; UKPP Cd1895, PTUC to Lawley, 8 October 1903, p.88; *The Age* (Melbourne), 4 August 1903, p.4; UKPP Cd1895, United T&LC of South Australia to CO, 10 November 1903, p.142, resolution against Chinese labour.
45. Vivian Bickford-Smith, E. Van Heyningen, & Nigel Worden, *Cape Town in the Twentieth Century: An Illustrated Social History* (1999), p.27.
46. Davenport, *Afrikaner*, p.249.
47. *The Times*, 9 September 1905; Katz, *Trade*, p.111.
48. *The Times*, 7 October 1905.
49. *Annual Report of the Trades Union Congress*, 1904, p.54, quoted in Zine Magubane, *Bringing the Empire Home: Race, Class, and Gender in Britain and Colonial South Africa* (2004), p.106.
50. *AEJ*, October 1903, p.8.
51. See Katz, p.111.
52. Quoted in Magubane, pp.106–107.
53. A. W., 'Yellow Slavery – and White!', *Westminster Review*, p.161, 5 May 1904, p.478, quoted in K. Harris PhD, pp.123–124.
54. Frank Hales, 'The Transvaal Labour Difficulties', *Fortnightly Review* (July 1904), p.111.
55. See Samuel, pp.458–459.
56. See Hyslop, 'Imperial', pp.398–421.
57. *Evening Post*, 8 February 1904, p.5, quoted in Jeremy Martens, 'Richard Seddon and Popular Opposition in New Zealand to the Introduction of Chinese Labour into the Transvaal, 1903–1904', *New Zealand Journal of History*, 42:42 (2008), p.177.
58. *The Advertiser* (Adelaide), 30 January 1904, p.7.
59. Ticktin, pp.71–72; Levy, p.202; UKPP Cd1895, African Labour League to Chamberlain, 30 July 1903, p.46; *SAN*, 3 October, 4 December 1903; MmP262, Shepherd to Merriman, 29 November 1903; Denoon, *Grand*, p.45. Their chairman was J. Quinn, a prominent baker (and also member of the White League). Other prominent members included W. W. Lorimer of the Shop Assistants' Union, R. J. Strickland, secretary of the South African Typographical Union, J. Reid of the Pretoria TLC and the Australian, R. L. Outhwaite, editor of the *South African Guardian*.
60. *Star*, 1 July 1903.
61. UKPP Cd1895, African Labour League to CO, 13 July 1903.
62. *SAN*, 28 December 1903.
63. See Chamberlain's comments confirming this dismissal of the colony in Julian Amery, *Life of Joseph Chamberlain, volume IV* (1951), p.338; see Headlam, Milner to Selborne, N.D., pp.550–558 for a similar expression from Milner.
64. MP10/293–5, Milner to Chamberlain, 13 July 1903.
65. Milner had publicly supported the plan and, so upset was Chamberlain, Milner offered to resign. For a full account of the incident, see Marlowe, *Milner*, pp.125–128; Davenport, *Afrikaner*, pp.238–239.
66. The Bond Party even changed their party manifesto in April, to mirror Merriman's words. See Davenport, *Afrikaner*, p.96.
67. Davenport, *Afrikaner*, p.189.

68. MmP15/24/2, Merriman to Richard Solomon, Transvaal Attorney General, 3 January 1904; see also Lewsen, *Correspondence*, Merriman to E. Sheppard, member of *Het Volk* and co-founder of South African Labour Party, 9 December 1903, p.414.
69. Anti-Asiatic Importation League, Letter *(Cape Town, January 1904)*.
70. H. E. S. Fremantle, 'The Political Position at the Cape', *Contemporary Review*, 84 (July/December 1903) 534.
71. Cronwright-Schreiner to Merriman, 13 April 1904, quoted in Ticktin, p.81.
72. *SAN*, 8 December 1903.
73. Mouton, S. E., '"A Free United South Africa under the union Jack": F S Malan, South Africanism and the British Empire, 1895–1924', 51:1 (May 2006), p.31.
74. *De Volksstem*, 4 July 1903, translated and quoted in Davenport, *Afrikaner*, p.255.
75. *Ons Land*, 1 December 1903; translated copy from CO48/573/202, Enclosure No. 6 in Despatch No. 832 of 2nd December 1903.
76. Hermann Giliomee, *The Afrikaners, Biography of a People* (London, 2003), p.6.
77. Davenport, *Afrikaner*, p.189.
78. See MmP23/245, C. H. Ockre[?] to Merriman, 19 November 1903; Elizabeth van Heyningen, 'Leander Starr Jameson', in Jane Carruthers (ed.), *The Jameson Raid, A Centennial Retrospective* (Houghton, SA, 1996), pp.181–191; Ian Colvin, *The Life of Jameson*, v.II, (1923), esp. p.212, 217; Lewsen, *Merriman*, p.267; Dubow, 'Imagining', pp.78–79
79. Andrew Thompson, 'Loyalism'.
80. See MP181/263, W. Thorne, Mayor of Cape Town, to Milner, 16 December 1903; response:pp.264–265, 17 December 1903; response:pp.266–267, 17 December 1903.
81. 'Speeches on The Chinese Question by Mr. Abe Bailey, MLA, Delivered in the House of Assembly, Cape Town, 8th and 16th March, 1904. Reprinted from Cape *Times* (Cape Town, 1904). Bailey and Merriman were friends before this issue but fell out over Chinese labour and never reconciled. My italics. See also Jameson to Sam Evans, 12 August 1903, quoted in Colvin, *vol.II*, p.220.
82. *Cape Times*, 28 November 1903.
83. CO48/572, Hely-Hutchinson to Chamberlain, 3 August 1903. See also J. W. Maquarries (ed.), *The Reminiscences of Sit Walter Stanford*, II, 1885–1929 (1962), p.242.
84. *Cape Times*, 28 November 1903.
85. UKPP Cd1941, Town Clerk, Port Elizabeth to Hely-Hutchinson, Cape, 14 January 1904, p.9.
86. UKPP Cd1941, Town Clerk, Aliwal North, to Hely-Hutchinson, 1 February 1904, p.64.
87. UKPP Cd1941, Rhodes: 26 December 1903, p.2; Woodstock: 4 January1904, p.4; Bedford:13 January 1904, p.35; Robinson: 4 February 1904, pp.64–65.
88. UKPP Cd1941, forwarded by Town Secretary to Prime Minister, Cape, 23 January 1904.
89. CO48/537/496–7, Secretary, Middelburg Town Council, to Cape Governor. 18 December 1903.

90. CO48/573/494–5, Town Clerk, Kimberley, to Cape Governor. 15 December 1903.
91. *SAN*, 15 December 1903.
92. UKPP Cd1941, Town Clerk, Cradock, to Hely-Hutchinson, 9 January 1904, p.6.
93. *Midland News*, 11 January 1904.
94. UKPP Cd1941, Secretary, Midden-Zwartland Branch of the Bond, to Hely-Hutchinson, 8 January 1904, p.7.
95. *SAN*, 19 January 1904; See also *SAN*, 8 December 1903, 21 December 1903; *Ons Land*, 8 December 1903.
96. Evans, et al., p.164.
97. Lewsen, *Correspondence*, Merriman to Solomon, 3 January 1904, pp.424–425.
98. *SAN*, 27 November 1903, my italics.
99. See MmP23/274, Harry Currey to Merriman, 5 December 1903; MmP23/304, J A Vosloo to Merriman. 17 December 1903.
100. 'Native Opinion' in Xhosa, published in King Williamstown in the Ciskei from 1884.
101. See *Imvo*, the newspaper he edited, on 30 November and 14 December 1903. See also L. D. Ngcongco, 'John Tengo Jabavu, 1859–1921', in Christopher Saunders (ed.), *Black Leaders in South African History* (1979), pp.142–156.
102. Lewsen, *Correspondence*, J. Tengo Jabavu to Merriman, 15 December 1903, p.415; see also Joyce F. Kirk, 'A "Native" Free State at Korsten: Challenge to Segregation in Port Elizabeth, South Africa, 1901–1905', *Journal of Southern African Studies*, 17:2 (June 1991), p.327.
103. MP191/203–5, F. Vere Stent to Milner, 26 March 1906; Lewsen, *Merriman*, p.276; J. H. Raynard, *Dr. A Abdurahman: A Biographical Memoir* (Cape Town, 1940); R.E. van Der Ross (ed.), *Say It Out: The A.P.O Presidential Addresses and other Major Political Speeches, 1906–1940, of Abdullah Abdurahman* (1990). He also started working with W. P. Schreiner for the first time over Chinese labour, an important partnership which fought against the colour bar.
104. Davenport, *Afrikaner*, p.249.
105. UKPP Cd1941, Sishuba to Hely-Hutchinson, 31 December 1903, p.2; December 1903, p.3; 7 January 1904, p.5; undated, pp.7–8. See also petitions in *South African Native Opinion*, 8 December 1903, *Mafeking Mail*, 6 November 1903, both quoted in Harris, PhD, p.127.
106. Cape *Hansard*, 2 July 1903.
107. CO48/527, Sprigg to Chamberlain, 3 August 1903.
108. CO48/537/330–1, Governor to Colonial Secretary, 9 December 1903; CO48/527/361, Colonial Secretary to Governor, 16 January 1903.
109. Davenport, *Afrikaner*, p.250; Lewsen, *Merriman*, p.20.
110. MP168/15/267, copy of telegram, Hely-Hutchinson to Lyttelton, 20 February 1904.
111. Lewsen, *Correspondence*, p.418 n78; See Conclusion; Duminy & Guest, p.333 and Lewsen, *Merriman*, p.270.
112. HE202/12, Jameson to J. F. Jones, 29 February 1904; Jameson to Jones, 29 February 1904; CO48/576, Governor to Colonial Secretary, 12 April 1904.
113. *Imvo*, December 1903, quoted in Warwick, p.172. See also André Odendaal, *Vukani Bantu! The Beginnings of Black Protest Politics in South Africa to*

	1912 (1984), pp.50–54; SNA188/76, copy of *South African Native Opinion*, 8 December 1903.
114. *Mafeking Mail*, 6 November 1903, in SNA185/2991/03, p.78.
115. See *SAN*, 8 December 1903; CO48/573, Cape Colony Correspondence, 1903.
116. UKPP Cd1941, Executive, SAPO (Thomas Milton, William Hunter, A. G. Francke), undated, p.5.
117. UKPP Cd1941, Hely-Hutchinson to Lyttelton, 20 February 1904.
118. *SAN*, 21 January 1904; see Harris, PhD, p.240.
119. *Indian Opinion*, 24 September 1903.
120. *Indian Opinion*, 14 June 1903.
121. *Indian Opinion*, 21 January 1904, 6 August 1903.
122. *Indian Opinion*, 11 June 1904.
123. MmP23/300, A. M. J. Helm to Julia Merriman, 15 December 1903.
124. G. Seymour Fort, *Dr. Jameson* (1908), p.254; see also Lewsen, *Correspondence, Diary*, 10 March 1904, p.419.
125. See, for instance, the coverage of a Cape Town meeting on 21 December 1903.
126. CO537/538, Secret report for January 1904, Cape Town, 3 February 1904, p.22.
127. UKPP Cd1895, Milner to Lyttelton, 3 January 1904, p.178; see similar sentiments in MmP23/289, Julia Merriman to Aggie Merriman, 11 December 1903.
128. See Mohamed Adhikari, 'Contending Approaches to Coloured Identity and the History of the Coloured People of South Africa', *History Compass* (September 2005), pp.1–16.
129. W. K. Hancock & J. van der Poel, *vol.II*, Botha/Smuts to L. Hobhouse, 13 June 1903, p.103. Written by Smuts, but signed by Botha, a common practice throughout the period.
130. MP171/170, Milner to Chamberlain, 17 February 1903.
131. J. A. Reeves, 'Chinese Labour in South Africa, 1901–1910' (MA thesis, University of the Witwatersrand, 1954), p.29.
132. *The Transvaal Leader*, 16, 18 January 1904.
133. *Rand Daily Mail*, 18 January 1904.
134. Hancock & van der Poel, vol. II, Circular, 14 December 1902, pp.58–59; see also Botha/Smuts to L. T. Hobhouse, 13 June 1903, pp.105–106.
135. Hancock & van der Poel, vol. II, Motion to Sir A. Lawley, 4 July 1903, pp.112, 116. My italics. See also Davenport, *Afrikaner*, p.256.
136. UKPP Cd1899, Milner to Lyttelton, 11 February 1904, p.22. See also his dismissal of Botha and Afrikaners generally in MP10/19/11, Milner to Lyttelton, 2 May 1904; CO48/572, Hely-Hutchinson to Chamberlain, 3 August 1903.
137. *Cape Times*, 8 December 1903; my italics.
138. *Hansard*, 21 March 1904, speech by Sir Gilbert Parker (Gravesend), p.305.
139. *Cape Daily Telegraph*, 19 January 1904.
140. MmP23/288, 11 December 1903, Smuts to Merriman
141. Reenen (ed.), Hobhouse to L. Hobhouse, 2 September 1903, pp. 280–281.
142. *Die Volkstem*, 6 January 1904, translation in Harris, PhD, pp.124–125.

143. Hancock. & van der Poel, vol. II, Smuts to Merriman, 15 February 1904, p.145; Botha/Smuts to L. Hobhouse, 13 June 1903, p.103.
144. Hancock & van der Poel, vol. II, Smuts to E. Hobhouse, 21 February 1904, pp.147–148; see also Smuts to E. Hobhouse, 28 February 1904, p.150.
145. UKPP Cd1895, Milner to Lyttelton, 3 January 1904, p.178.
146. *Bloemfontein Post*, 2 April 1903; pp.9–11 November 1904.
147. MmP25/205, Steyn to Merriman, 10 November 1905.
148. *The Friend*, 12 January 1904.
149. *The Friend/Dr Vriend*, 18 January 1904.
150. *Bloemfontein Post*, 2 April 1903.
151. *Bloemfontein Post*, 9 November 1904.
152. *De Volksstem*, 16 February 1904.
153. CO291/67/259, C. G. Radloff to Chamberlain, 19 September 1903.
154. Anonymous, *British Indians in the Transvaal and the Orange River Colonies*, Reprinted from *India*, 17 October 1902.
155. The *Sydney Morning Herald*, 11 January 1904, p.7; see also *The New Zealand Herald, 100 Years Ago*, 23 February 2004.
156. The *Argus*, 12 January 1904, p.5; Martens, 'Seddon', 185. He also wrote to many local organisations, such as the Cape Anti-Asiatic League. See *The Advertiser* (Adelaide), 30 January 1904, p.7.
157. The *Advertiser*, 21 January 1904, p.4 and 6 August 1903,p.6; *The Sydney Morning Herald*, 7 March 1904, p.6; *The Argus*, 19 March 1904; see also UKPP Cd1895, United T&LC of South Australia to CO, 10 November 1903, p.142, resolution against Chinese labour.
158. UKPP Cd1941, Premier, New Zealand to Colonial Secretary; Prime Minister, Australia to Colonial Secretary, p.28; UKPP Cd1895, Ranfurly to Lyttelton, 20 January 1904, p.231.
159. MP178/789, Milner to Earl Grey, Governor-General, Canada, 14 March 1904.
160. Huttenback, 'British', pp.129–130, 132–135; MmP24/79, Goldwin Smith to Merriman, 28 February 1904; CO50/566, Grey to Lyttelton, 18 February 1904; CO50/544, Laurier to Lyttelton, 20 February 1904; *The Advertiser*, 15 January 1904, p.5.
161. *The Argus*, 13 January 1904, p.4; House of Representatives speech, Conroy (NSW), 19 March 1904, p.15.
162. *Bulawayo Observer*, 11 April 1903.
163. Huttenback, *Gandhi*, p.340. See CO417/381 for more information on the Rhodesian plans.
164. *Times of Natal*, 20 January 1904.
165. Philip Harrison, 'Reconstruction and Planning in the Aftermath of the Anglo-Boer South African War: The Experience of the Colony of Natal, 1900–1910', *Planning Perspectives*, 17 (2002), p.167.
166. UKPP Cd1895, Milner to Lyttelton, 3 January 1904, p.178. See also MP179/72, Maydon to Milner, 11 December 1902; response: 179/133–4, 22 December 1902; Cd1899, Milner to Lyttelton, 6 February 1904; Cd1941, Milner to Lyttelton, 15 February 1904, p.22; MP178/178, Hine to Milner, 27 March 1904; MP179/72, Maydon to Milner, 11 December 1902; MP179/73, Maydon to Milner, 5 March 1904; R. Jebb, *Studies in Colonial Nationalism* (London, 1905), p.130.

Notes 215

167. John Lambert, ' "The Thinking is done in London": South Africa's English Language Press and Imperialism', in C. Kaul (ed), *Media and the British Empire* (2006), p.47.
168. MP176/264, Collins to Milner, 29 August 1904.
169. *The Advertiser*, 14 July 1903, p.5; 15 January 1904, p. 5.
170. *The Advertiser*, Tuesday 19 January 1904, p. 5.
171. *Ilanga lase Natal*, 1903, quoted in Warwick, p.172.
172. LtG101/11, Maritzburg Chamber of Commerce to Milner, 10 March 1904.
173. UKPP Cd1986, Acting Governor, Sir H. Bale, Natal, to Lyttelton, p.3.
174. UKPP Cd1941, Legislative Council, Labour Importation Draft Ordinance, Second Reading, 20 January 1904, p.61; Kinloch Cooke, p.3.
175. UKPP Cd1941, Lyttelton to Premier, New Zealand and Prime Minister, Australia, 20 January 1904, p.29.
176. UKPP Cd1895, Chamber of Mines meeting, 2 December 1903, p.144; UKPP Cd1941, Legislative Council, Labour Importation Draft Ordinance, Second Reading, 20 January 1904, pp.58–59.
177. P.A. Molteno, *A Federal South Africa* (1896), p.viii.
178. British Colonist, 'British Rule in the Transvaal', *Contemporary Review*, 85 (January/June 1904), p.329, 351.
179. UKPP Cd1941, Legislative Council, Labour Importation Draft Ordinance, Second Reading, 20 January 1904, p.61.
180. UKPP Cd1941, S. P. Coetzee Jnr. to Lieutenant-Governor, 4 January 1904, p.14. Of the 50–60 present, 5 voted against.
181. *Star*, 5 December 1903; *SAN*, 16 January 1904; Katz, pp.112.
182. *The Argus*, 19 March 1904, p.15; *Melbourne*, 22 March 1904, p.5; *The Brisbane Courier*, 1 April 1904, p.4.
183. *The Aborigines' Friend*, February 1904, pp.2,11.
184. CS407/11033/03, 26 December 1903, Marais Rex Esqu., Pretoria, to Colonial Secretary.
185. *Hansard*, 21 March 1904, speech by H. Campbell-Bannerman, pp.252–253, 271.
186. *Hansard*, 21 March 1904, p.212.
187. *Hansard*, 21 March 1904, speech by Sir Frederick Milner (Nottinghamshire, Bassetlaw).
188. *The Brisbane Courier*, 22 January 1904, p.5; *The Argus*, 22 January 1904.
189. *Hansard*, 11 February 1904, speech by the Duke of Marlborough, p.997
190. *Hansard*, 21 March 1904, speech by The First Commissioner of Works (Lord Windsor), p.174.
191. *Hansard*, 21 March 1904, speech by Lord Norton, p.126.
192. *Hansard*, 21 March 1904, speech by the Secretary of State for Foreign Affairs (The Marquess of Lansdowne), pp.206–207.
193. Katz, *Trade*, p.121. UKPP Cd1896, Minority Report, Appendix, Mr. Terbutt's Letter, 3 July 1902, p.65; *Transvaal Legislative Debates*, 28 December 1903, p.75; *Hansard*, 21 March 1904, speech by H. Campbell-Bannerman, p.261.
194. *Transvaal Leader*, 2 April 1903; see also *The Advertiser*, 'To the Editor', 18 February 1904, p.9.
195. Fremantle, 'Political Position', p.539.
196. *The New Zealand Herald*, 23 February 1904. See also *The Argus*, 22 March 1904, p.5.

197. CO291/95, Selborne to Elgin, 6 January 1906. See also 9 and 15 January.
198. J. A. Hobson, 'Capitalism and Imperialism in South Africa', *Contemporary Review*, 77 (January/June 1900), pp.1–17; *The Psychology of Jingoism* (1901); *Imperialism: A Study* (1902); P. Cain, 'British Radicalism, the South African Crisis, and the Origins of the Theory of Financial Imperialism', in Omissi & Thompson (eds.), pp.173–193.
199. Cain, p.179. For more on the importance of the South African War in Britain and debates about empire, see Thompson, 'Language', especially ft.14.
200. *South African News*, 22 June 1905; W. Wybergh, 'The Transvaal and the New Government', *Contemporary Review* (March 1906), p.316; Speech by Whiteside, *Debates of the Transvaal Legislative Assembly*, 19 June 1907, p.173; E. Se. Beesly, 'Yellow Labour', The Positivist Review, 1 April 1904.
201. *The Star*, 17 January 1903. Report of Chamberlain's speech at a Wanderers' Rink banquet; see also quoted in Amery, *Joseph*, p.332.
202. For a brief analysis of Chamberlain's industrial links, see R. V. Kubicek, 'Economic Power at the Periphery: Canada, Australia and South Africa, 1850–1914', in Raymond E. Dumett (ed.), *Gentlemanly Capitalism and British Imperialism, The New Debate on Empire* (1999), pp.113–126. See complaints about the loan in UKPP Cd1895, Enclosure 2 in No.28, Motion-A Kock, seconded by P J Nys, re the War Debt, 8 January 1903, p.53.
203. Katzenellenbogen, *South Africa and Southern Mozambique: Labour, Railways and Trade in the Making of a Relationship* (1982), p.65
204. See Gerald Shaw, *The Cape Times*, Chapter.1; Porter, 'Milner', pp.332–333; Beaumont, pp.177–194; Potter, *News*, pp.99–102. The *Rand Daily Mail* was later bought by the mine magnate, Abe Bailey. See Hancock & van der Poel (eds.), E. P. Solomon to J C Smuts, 26 April 1905, p.267.
205. Potter, *News*, pp.41, 43, 45.
206. Potter, News, pp.100, 119.
207. Porter, 'Milner', p.325; Gerald Shaw, *The Cape Times*, p.13.
208. Beaumont, pp.177–194. Once he resigned, the newspaper hired G. G. Robinson, formerly Milner's private secretary, as editor and it then supported importation, leading to accusations of undo press influence.
209. Ticktin, p.70.
210. William Mather, *Chinese Workers on the Witwatersrand Mines* (Johannesburg, 1904), p.25; MP341/12, ISAA pamphlet: 'Asiatic Labour in the Transvaal', n.d.; *The Argus*, Melbourne, 22 March 1904, p.5.
211. See Katz, *Trade*, p.115. He was not paid by Corner House; other mining records are unavailable. His deceit was not uncovered until Chinese labour had been approved.
212. *Star*, 16 October 1903; Katz, *Trade*, p.113; UKPP Cd1897, p.1; *SAN*, 4 December 1903.
213. Strong. He claimed that his reason for leaving were because of his opposition to strikes.
214. *SAN*, 12 August 1903. See also *Thomas Burt, Northumberland Miners' Mutual Confident Association. 'Monthly Circular'*. April 1904. 'Chinese Labour: How Public Opinion is Ascertained, or Manufactured in the Transvaal.'
215. *SAN*, 17 February 1904; see also *SAN*, 21 December 1903; *Cape Times*, 21 December 1903.

216. See the petitions in UKPP Cd1895, Lawley to Chamberlain, 22 August 1903, p.57; Lawley to Chamberlain, 31 August 1903; James B. MacDonald, Secretary, CMLIA, to Lawley, 29 August 1903, pp.60–61.
217. UKPP Cd1895, WT&LC to CO, 31 August 1903, p.55; LtG144/127/195, Botha to Currie, 28 March 1906; Cd1941, Transvaal Legislative Council Labour Importation Ordinance Debates, 19 January 1904, p.63; Cd1899, Milner to Lyttelton, 29 January 1904, p.4; Milner to Lyttelton, 30 January 1904, pp.4–5.
218. *De Volksstem* 27 January 1904, 30 December 1903, p.5, 30 January 1904, p.6
219. H. O'Kelly Webber, *The Grip of Gold. The Life Story of a Dominion* (London, 1936), pp.133–134.
220. HE202/12, Chaplin to Reyersbach, Eckstein & Co., 26 November 1904.
221. UKPP Cd1895, Lawley to Lyttelton, 7 December 1903, p.127.
222. Fraser, M. & A. Jeeves, *All That Glittered: Selected Correspondence of Lionel Phillips, 1890–1924* (1977), Phillips to L. Reyersbach, Strictly Private, 13 November 1903, p.121.
223. R. J. Pakeman, *Political Letters from the Transvaal* (Johannesburg, 1904).
224. J. Russell, *Liberal Landslide. The General Election of 1906* (1973), p.113.
225. Wheatcroft, *Randlords*, p.222.
226. UKPP Cd1895, Lyttelton to Milner, 18 February 1904, p.22.
227. See *St. James Gazette*, 11 January 1904; *Pall Mall Gazette*, 11 January 1904.
228. UKPP Cd2104, Lyttelton to Gov-General Lord Northcote (Australia), 20 May 1904, p.3.
229. See also A. E. Munro, *Transvaal Chinese Labour Difficulty*, p.128; Bruce, p.90; *The Times*, Letters to Editor, from Alfred Hillier, 15 February 1906, p.7; L. S. Amery, 22 February 1906, p.7; John Francis Molyneux, Health Officer, Chifu, 4 June 1906, p.6.
230. *SAN*, 8 December 1903.

4 A Question of Honour: Slavery, Sovereignty, and the Legal Framework

1. Richardson, *Chinese*, p.29.
2. See coverage in the *Aborigines' Friend* between 1904 and 1907.
3. See, for Britain, Catherine Mills, 'The Emergence of Statutory Hygiene Precautions in the British Mining Industries, 1890–1914', *The Historical Journal*, 51:1 (2008), pp.145–168.
4. A discourse which has continued to overshadow the historiography on the subject, particularly in Tinkler and, more recently in South Africa, Karen Harris, 'Sugar and Gold: Indentured Indian and Chinese Labour in South Africa', *Journal of Social Science*, 25 (2010), pp.147–158.
5. Northrup, pp.110–111; Robert Irick, *Ch'ing Policy Toward the Coolie Trade 1847–1878* (1982).
6. Grant, pp.106–107.
7. Cain, 'British Radicalism', p.175.
8. Russell, p.106.
9. David Feldman, 'Jews and the British Empire c.1900', *History Workshop Journal*, 63 (2007), 75.

10. Herbert Samuel, MP, 'The Chinese Labour Question', *Contemporary Review*, 85 (January/June 1904), 465.
11. H. C. Thomson, 'Chinese Labour and Imperial Responsibility', *Contemporary Review*, 89 (January/June 1906), 437.
12. John Burns, MP, 'Bondage for Black. Slavery for Yellow Labour.' Reprinted from *Independent Review*, May 1904, (London).
13. William Harcourt, MP, 'Native Labour in South Africa. Two Letters to the Times', *Reprinted by the Liberal Publication Dept. (1903): Letter 1*, 6 February 1903.
14. Hansard, HC Debate 21 March 1904, Chinese Labour for the Transvaal-vote of censure, speech by H. Campbell-Bannerman, pp.262–263.
15. Hansard, 21 March 1904, speech by Moutlon (Launceston, Cornwall), p.289.
16. Hansard, 21 March 1904, speech by Moutlon, p.299.
17. Hansard, 21 March 1904, speech by the Lord Bishop of Hereford, p.120.
18. Thomas M'Kinnon Wood, Liberal candidate for St. Rollox (Glasgow), in *Glasgow Herald*, 5 January 1906, 6, quoted in Scott C. Spencer, 'British Liberty Stained: "Chinese Slavery," Imperial Rhetoric, and the 1906 British General Election', *Madison Historical Review*, 7 (May 2010), 13.
19. George Barlow, 'Chinese Labour in South Africa', *1904 sonnet in The Poetical Works, 1902–1914*. The most reproduced newspaper cartoon was from the *Daily Mirror*, 1 August 1905.
20. John Burns, 'Slavery in South Africa', *Independent Review*, 2 (May 1904), 606. This was later turned into a pamphlet for the 1906 election campaign.
21. Grant, p.101.
22. John Burns, *Bondage for Black. Slavery for Yellow Labour, Reprinted from Independent Review* (May 1904) (London).
23. Hyslop, 'Imperial Working Class', 414.
24. *De Volksstem*, 27 April 1904, quoted in Harris PhD, p.124; *The Adelaide Advertiser*, 29 March 1904, p.5; *Sydney Morning Herald*, 30 March 1904, p.10.
25. Edward B. Rose, *Uncle Tom's Cabin*, (London, 1906), p.12, Rose was the late President of the Witwatersrand Mine Employee's & Mechanics' Union. The pamphlet was published by the South Africa Free Press Committee for a British audience.
26. Stoler & Cooper, p.37.
27. John A. Garrard, *The English and Immigration, 1880–1910* (1971), pp.90–91; Jill Pellew, 'The Home Office and the Aliens Act, 1905', *The Historical Journal*, 32:2 (June 1989), 374; for tariff reform debates, see A. Howe, *Free Trade and Liberal England, 1846–1946* (1997).
28. Spencer, p.20.
29. See the *Star*, 27 January & 4 February 1904, Legislative Council meetings.
30. *North-China Herald & S.C. & C. Gazette*, 20 May 1904, Letter by Diogenes.
31. *Hansard*, 21 March 1904, speech by Lord Monkswell, p.182.
32. *Hansard*, 11 February 1904, speech by the Archbishop of Canterbury, p.103; see also speech on 21 March 1904.
33. LtG159, Furse to Milner, 16 February 1904.
34. UKPP Cd2026, Milner to Lyttelton, 3 March 1904, pp.1–2. See also *Hansard*, 21 March 1904, speech by the Earl of Selborne, pp.147–155.
35. HE72, W. Beit to Eckstein & Co, 19 February 1904.

36. FLD 9/147/31, Chinese Labour for the Rand, 5 April 1905, Contract of Labour, pp.3–4.
37. CO291/69/15, W D Evans to [?], 7 March 1904.
38. UKPP Cd2401, Milner to Lyttelton, 28 November 1904, attached report, pp.42–44.
39. Cd2401, Milner to Lyttelton, 17 December1904, p.42; General Report on Chinese Labour by Samuel Evans, 13 February 1905, pp.80–85.
40. UKPP Cd2026, Lyttelton to Milner, 7 May 1904, p.49.
41. Smith, 'Capitalism and the War', p.71.
42. FLD9/147/30/1A, Evan's report on experiment to date, November 1904.
43. Denoon, *Grand*, p.207.
44. HE88/38/390. Eckstein to Wernher, 7 March 1904.
45. FLD176/36/1, FLD to CMLIA, 5 September 1905.
46. FLD9/147/35/1, Memorandum on Chinese Labour, pp.5–6.
47. John H Harris, *Coolie Labour in the British Crown Colonies and Protectorates* (London, 1910), p.17, produced by the Anti-Slavery & Aborigines Protection Society.
48. FLD9/147/35/1, Memorandum on Chinese Labour, FLD, undated, p.3; HE88/165, Eckstein to Wernher, 19 October 1903.
49. MmP24/17, Solomon to Merriman, 11 January 1904, my italics.
50. *Hansard*, 21 March 1904, speech by Lord Stanley of Alderley, p.146
51. See Chapters 5 and 6 for more on the implementation of the scheme.
52. Cd1945, Article V of the Treaty of 1860 between China and Great Britain, p.1; see also CO879/85/755, Waiwupu to the Superintendents of Northern and Southern Trade, *Tientsin Official Gazette*, 22 March 1904, enclosed in FO to CO, 12 May 1904.
53. The Department of Foreign Affairs, set up in July 1902, which took precedent over the other six Ministries of State as part of the peace agreement between China and European powers after the Boxer Rebellion.
54. CO879/85/755, Chinese Waiwupu to Superintendents, Northern & Southern Trade, 22 March 1904.
55. Richardson, *Chinese*, p.224.
56. Captain R. A. Crouch MP, 'An Australian View of the War', *Contemporary Review*, 86 (July/December 1904), 180.
57. Cd1945, Chang Ta-Jên, Chinese Minister, London, to Lansdowne, Foreign Secretary, 11 February 1904, p.3.
58. CO291/75/7860, Minutes of Meeting between Chinese and British Governments re: Engagement of Chinese Labourers, 24 February 1904.
59. For general information, see Immanuel C. Y. Hsü, *The Rise of Modern China*, 4th edition (1990), pp.355–375. For more on South African Chinese connections, see Yap & Man, pp.92–93.
60. CO291/67/20153, Chinese grievances Petition from Chinese community of December 1902, 25 May 1903; see also Harris, PhD, p.277.
61. Yap & Man, p.93.
62. Cd1945, Chang Ta-Jên, Chinese Minister, London, to Lansdowne, Foreign Secretary, 11 February 1904, p.3.
63. Cd2026, Milner to Lyttelton, 29 March 1904, p.9.
64. Cd2026, CO to FO, 5 March 1904, pp.3–4.
65. Cd2026, Lyttelton to Milner, 4 April 1904, p.11.

66. FLD130/19/3, Consular Matters, quoted in Yap & Man, p.111.
67. *Star*, 9 May 1905.
68. Cd1945, Chang Ta-Jên to Lansdowne, 11 February 1904, pp.3–4; Ching-Hwang, p.341.
69. CO291/75/7860, Minutes of Meeting between Chinese and British Governments re Engagement of Chinese Labourers, 24 February 1904.
70. UKPP Cd2026, Lyttelton to Milner, 29 April 1904, pp.45–46; Cd2026, No.42, Lyttelton to Milner, 3 May 1904, p.46.
71. FLD7/147/20/1, Extract from Letter from Mr. Perry and Mr. Hamilton, 16 February 1904.
72. UKPP Cd2026, Lyttelton to Milner, 29 April 1904, pp.45–46; Lyttelton to Milner, 3 May 1904, p.46; Milner to Lyttelton, 2 May 1904, p.46; Lyttelton to Milner, 5 May 1904, p.47; Milner to Lyttelton, 16 May 1904, p.52.
73. FLD25/60/06, Solomon to Jamieson, 23 March 1906.
74. FO228/215/2, H. Bertram Cox, CO, to FO, 5 March 1904.
75. MP42/3/179, Milner to Selborne, 23 October 1905, p.559.
76. CO67/106/9614, Minute for all departments, 27 April 1897, quoted in Robert V. Kubicek, *The Administration of Imperialism: Joseph Chamberlain at the Colonial Office* (1969), pp.34–35.
77. Business diary, 16 January 1903, quoted in Amery, *Chamberlain*, p.333.
78. UKPP Cd1897, Labour Commission Evidence, p.608.
79. H. C. Thomson, 'Chinese Labour and Imperial Responsibility', *Contemporary Review*, 89 (1906:January/June), pp.435–436.
80. FLD225/62/55, A S H Cooper to Selborne regarding allegations by Mr Cooper whilst in Gaol, 8 February 1906; FLD50/6/1, 3 March 1909, Acting Supt., R I Purdon, to J de Villiers Roos Esq., Acting Secretary to the Law Department.
81. See FLD269.
82. HE73/128–9, Wernher to Eckstein, 30 September 1904; HE250/139, Goodwin, Acting General Manager, Rand Mines Ltd, to Eckstein, 5 December 1899.
83. HE221/47c, Schumacher to Wernher, 9 May 1904.
84. FLD9/147/35/1, Memorandum on Chinese Labour, p.7; HE88/144, Eckstein to Wernher, 12 October 03; HE221/47c, Articles of Association of the Chamber of Mines Labour Importation Agency, Limited, April 1904.
85. FLD145/27/50. While the specific identity of Brazier is unknown, it is likely he was part of the well-established British commercial family in China of the same name, making him ideal to negotiate this new commercial activity.
86. FLD9/147/31, Chinese Labour for the Rand, 5 April 1905, Contract of Labour, p.1.
87. FLD *Annual Report*, 1905–1906, Appendix 8; FLD9/147/35/1, Memorandum on Chinese Labour in the Transvaal, pp.5–6.
88. UKPP Cd2026, Instructions for Transvaal Emigration Agents in China, from Ordinance, pp.40–42.
89. UKPP Cd2026, Instructions, pp.40–42.
90. FLD9/147/35/1, Extract from Amoy Intelligence Report, May–July 1904, pp.34. See T. G. Otte, *The China Question: Great Power Rivalry and British*

Isolation, 1894–1905 (2006), pp.268–325, 328 for more on British competition with Japan and Germany in the region.
91. *The Straits Times*, 26 May 1904, p.5.
92. MP180/96, Perry to Milner, 8 July 1904.
93. FLD9/147/35/1, Extract from Amoy Intelligence Report, May–July 1904, p.2.
94. FLD9/147/35/1, Perry to Bagot, 3 October 1904.
95. Trew, p.102.
96. FLD9/147/30/1B, Report, S/S Indravelli, undated. See also *The Hong Kong Telegraph*, 1 September 1904; FLD9/147/31, Arnold Forster, London Mission, Wuchang, China, to Henry Fowler MP, 1904.
97. FLD9/147/35/1, M. Nathan, Governor, Hong Kong, to Lyttelton, 31 October 1904.
98. FLD131/20/3, F. Perry to Chairman and Members and Committee of Groups, Chamber of Mines, 30 July 1904.
99. HE72/371–3, W. Beit to Eckstein & Co., 4 March 1904.
100. J. Fairbank et al. (eds.), *The I. G. In Peking, Letters of Robert Hart, volume II* (1975), Hart to Campbell, 24 April 1904.
101. FLD9/147/30/1B, SS Tweedale, by Superintendent., FLD, undated but in July 1904.
102. FLD83/11, Superintendent, FLD, to TEA, Hong Kong, 27 October 1904.
103. FLD9/147/35/1, Superintendent, FLD, to TEA, Hong Kong, 19 October 1904.
104. FLD9/147/35/1, Memorandum on Chinese Labour, p.9.
105. FLD12/147/41/1, Order for the Return of a Labourer to his Country of Origin, 22 February 1905.
106. FLD4/147/7/1, Perry to Farrar, 28 May 1904.
107. HE221/49, Walter L Bagot, General Manager, CMLIA, to Perry, Future organisation in China, from Board meeting of CMLIA, undated, 1905. See also HE73/99, Wernher to Eckstein, 9 September 1904.
108. FLD184/39, Lyttelton to Milner, 31 March 1905, forwarding an extract from FO despatch by A. J. Flaherty, Acting Vice-Consul at Tientsin, 16 December 1904.
109. Richardson, 'Chinese', p.272.
110. Richardson, 'Coolies', p.177.
111. Richardson, *Chinese*, pp.118, 104, 114; Ian Phimister, 'Foreign Devils, Finance and Informal Empire: Britain and China c.1900–1912', *Modern Asian Studies*, 40:3 (2006), 741, 745, 752–753. See Table 4.1.
112. Cd1895, Report of H. Ross Skinner, p.80.
113. FLD184/39, 31 March 1905, Lyttelton to Milner, forwarding an extract from FO despatch by A. J. Flaherty, Acting Vice-Consul, Tientsin, 16 December 1904.
114. FLD83/11, Report on a Tour through Shantung Province (with special reference to S. African emigration.), by E. D. C. Wolfe, TEA, Shantung, 31 July 1905.
115. FLD184/39, TEA, Chinwangtao, to Superintendent, FLD, 18 April 1905.
116. FLD184/39, Advances and Allotments, Official Chamber of Mines view from de Jongh, President, to Malcolm, Governor's Office, 1 October 1906.
117. FLD131/20/1, Fairfax, Inspector, FLD, to Superintendent, interviewing Chinese miners on Jumpers Deep Ltd, 23 June 1905.

222 Notes

118. FLD131/20/2, Superintendent, FLD to Lieutenant-Governor, 30 June 1905. See Table 4.1.
119. FLD131/20/8, Brazier to General Manager, CMLIA, 4 October 1906; see also FLD133/21/6, Cable 10, 13 April 1904.
120. Brazier to General Manager, CMLIA, 4 October 1906.
121. FLD133/21/6, English Version of Notice Issued by Recruiting Agents in Hong Kong, 13 April 1904, p.3.
122. FLD9/147/31, forwarded Contract of Labour with commentary by unspecified agent, 5 April 1905, p.5.
123. T. Burt, *A Visit to the Transvaal* (Newcastle-upon-Tyne, 1905), p.61.
124. FLD24/25/06, Horst to E. C. Mayers, FLD, 29 January 1906.
125. FLD9/147/30/1A, Arrival Report, SS Tweedale, by Superintendent, FLD, undated.
126. *The Hong Kong Telegraph*, 16 September 1904, gave a favourable description, as did Rev. T. W. Pearce, 'Chinese Coolie Labour in South Africa, A Statement from the Chinese Missionary Standpoint', *Chronicle of the London Missionary Society* (1904), pp.226–228; Rev. Arnold Foster, 'Chinese Coolie Labour in South Africa' (1905), pp.19–23, describes the recruitment methods as misleading and equivalent to slavery, as does his correspondence with the CO in FLD9/147/31.
127. Rose, p.12.

5 Sex, Violence, and the Chinese: The 1905–06 Moral Panic

1. The Johannesburg *Star*, 23 March 1905. See also *Bloemfontein Post*, Special issue, Chamber of Mines meeting, 21 October 1904.
2. *The Daily News*, 9 January 1906, from a series of highly critical articles by H. W. Massingham on Chinese labour conditions.
3. FLD14/147/52, FLD Convictions and Sentences of Chinese.
4. See the traditional definition of 'moral panics' in S. Cohen, *Folk Devils and Moral Panic* (1972), p.9.
5. Thompson, *Moral*, p.17.
6. Gary F. Jensen, *The Path of the Devil: Early Modern Witch Hunts* (2007).
7. Chambliss & Mankoff 1976: pp.15–16, quoted in Thompson, *Moral*, p.18; see also Frank Furedi, *Culture of Fear: Risk Taking and the Morality of Low Expectation* (2002).
8. Dane Kennedy, *Islands of White: Settler Society and Culture in Kenya and Southern Rhodesia, 1890–1939* (1987), pp.138–147.
9. Kennedy, pp.145, 146, 188, referenced in McCulloch, p.8.
10. Anderson, pp.67–68.
11. McCulloch, p.9.
12. Ellis, 'Violence and History', p.469.
13. Anderson, p.51.
14. Cornwell, pp.441–442.
15. See Frantz Fanon, *The Wretched of the Earth* (1963).
16. C. van Onselen, 'The Witches of Suburbia. Domestic Service on the Witwatersrand, 1890–1914', in *Studies in the Social and Economic History of*

the *Witwatersrand, 1886–1914,* vol2, *new Nineveh* (1982), pp.50–53; Jeremy C. Martens, 'Settler Homes, Manhood and "Houseboys": An Analysis of Natal's Rape Scare of 1886', *Journal of Southern African Studies,* 28:2 (June 2002), 379.
17. N. Etherington, 'Natal's Black Rape Scares of the 1870s', *Journal of Southern African Studies* 15 (1988), 43.
18. See Tables, pp.1–4.
19. Chamber of Mines Annual Report, 1905, p.44.
20. Arthur Mawby, *Gold Mining volume II,* pp.637–638, 714; *Star,* 28 March 1906.
21. *Transvaal Advertiser,* 19 July 1906.
22. *Star,* 16 March 1905. See also FLD133/147/65/1, Selborne to Lyttelton, 12 June 1905, for official concerns.
23. FLD133/147/65/2, WTLC to R. Solomon, undated, concerning 11 August 1906 meeting re: Simmer East. See also James Bridgeman, quoted by Katz, *Trade,* p.128.
24. *Daily News,* 2 July 1906; *Manchester Guardian,* 10 May 1906; Sacks, p.90.
25. See Balla Pillay, *British Indians in the Transvaal. Trade, Politics, and Imperial Relations, 1865–1906* (1976), p.199; CO291/82, Selborne to Lyttelton, 22 May 1905.
26. *Indian Opinion,* 15 July 1905.
27. Anderson, p.50
28. Balfour-Browne, *South Africa* (1905), p.190.
29. FLD51/6/8, Riots and Disturbances, Manager, South Nourse Ltd, to Secretary, FLD, 30 May 1905.
30. FLD51/6/15, Riots and Disturbances, Horst, Manager, New Kleinfontein Co Ltd, to Superintendent, FLD, 23 September 1905.
31. Trew, p.117.
32. FLD10/147/36/19, Acting Superintendent, FLD, to Lieutenant-Governor, 16 May 1905.
33. *Indian Opinion,* 6 January 1906, letter to editor, M. Berthasre Maharaj, 29 December 1905, p.4
34. *Indian Opinion,* 23 September 1905, p.638
35. FLD25/49/06, Wearing of disused Army Uniforms by Chinese Coolies, John Cochran, Sapper Royal Engineers, to Staff Captain, Pretoria Sub-Division, 21 February 1906; see also T. E. Stephenson, Major-General, Commanding Transvaal District, to Selborne, 22 February 1906
36. FLD25/49/06, John Cochran, Sapper Royal Engineers, to Staff Captain, Pretoria Sub-Division, 21 February 1906.
37. See, for example, a photograph of 'Chinese Police, Geldenhuis Deep', taken by J. Y. Simpson, 1905, from ALBUM 59 (4044), at the Library of South Africa, Cape Town.
38. Captain A. A. S. Barnes was among the primary group of British officers sent to raise and organise the Chinese Regiment at Weihaiwei in 1899, serving in it during the Boxer Uprising in 1900, and was with them for seven years in total. In August 1904 he was assigned to Chefoo as the Transvaal Government Agent for Emigration, where it seems he recruited many soldiers who had previously served under him. Certainly, at his wedding in March 1905, there were drummers and buglers from the regiment in full costume. See Captain A. A. S. Barnes, *On Active Service with the Chinese Regiment. A Record*

of the Operations of the First Chinese Regiment in North China from March to October 1900 (London, 1902); The North China Herald and S.C. & C. Gazette, 17 March 1905, p.530.
39. Image 3: Postcard: On the Rand: Chinese Policemen, Braune & Levy, Johannesburg.
40. Transvaal Advertiser, 12 July 1906.
41. Cd2786, Selborne to Lyttelton, 28 October 1905, p.45; Star, 3 July, 17–19 August 1905.
42. The Minneapolis Journal, 5 November 1905, First News Section, p.3; see also the New Zealand Grey River Argus, 20 November 1905, p.4; London Daily Mail, 25 September 1905, p.7; The N. C. Herald and S. C. & C. Gazette, 13 October 1905, p.57.
43. The Mercury (Tasmania), 6 September 1905, p.5; see also The Register (South Australia), 5 October 1905.
44. See The Register, 'Transvaal Colony. Another Chinese Trouble', 11 April 1905, p.5.
45. FLD16/147/81/10, Outrages, Heidelberg, 18 April 1906; Lieutenant-Governor's response, 23 April 1906; two accounts from local SAC on attacks in area, April 1906.
46. McCulloch, p.21.
47. Anderson, p.50.
48. Star, 13, 14 February 1905.
49. FLD16/147/81/2, Outrages, Deputation of Delegates from Outlying Districts, Pretoria, 6 September 1905, Report of Proceedings, Deputation to his Excellency the Lieutenant-Governor on the Question of Desertion of Chinese Labourers from the Mine. See similar complaints in FLD16/147/81/7, Outrages, Complaints by Ellewee, Davidson, Schoeman and Others to Botha, 13 January 1906, response from SAC, 8 February 1906.
50. FLD16/147/81/2, Outrages, Deputation of Delegates from Outlying Districts, Pretoria, 6 September 1905, Report of Proceedings, Deputation to his Excellency the Lieutenant-Governor on the Question of Desertion of Chinese Labourers from the Mine. See similar complaints in FLD16/147/81/7, Outrages, Complaints by Ellewee, Davidson, Schoeman and Others to Botha, 13 January 1906, response from SAC, 8 February 1906.
51. See Chapter 3, 'Corruption'.
52. Reprinted in The Sydney Morning Herald (NSW), 1 December 1905, p.7.
53. Phillips, Transvaal, p.138; Star, 20 April 1905, p.6; Schumacher, pp.15–16.
54. Selborne to Churchill, 8 April 1906, quoted in R. Churchill, pp.536–537.
55. Martin & Hyam, Reappraisals, p.184; see also Hyslop, Bain, p.165.
56. Morning Post (Queensland), 17 May 1906, p.2.
57. Hancock, Smuts Papers, vol. II, translation of letter from M. T. Steyn, 27 May 1906, p.279; see also translation of letter, M. T. Steyn to Smuts, 12 April 1906, p.258.
58. A. E. Polley, John Chinaman on the Rand. A New Form of Torture (London, 1905), p.82.
59. FLD16/147/81/3, 14 July 1905, Charles A O Bain to Sect., Rand Pioneers.
60. FLD16/147/81/3, 14 July 1905, Charles A O Bain to Sect., Rand Pioneers.

Notes 225

61. FLD16/147/81/3, 22 August 1905, Lieut Colonel, Chief Staff Officer, SAC, to Lieutenant-Governor; see coverage in the *Daily Express*, 1 September 1905.
62. FLD50/6/1, 12 and 18 April 1906.
63. SP56/1-5, Correspondence with various colonial governors and their secretaries, C. P. Crewe to Selborne, 13 September 1905.
64. SP56/7-8, Selborne to C. P. Crewe, 19 September 1905.
65. Krikler, 'Social Neurosis', p.67.
66. Krikler, 'Social Neurosis', p.70.
67. Lagden to Saunders, 16 June 1904, quoted in Krikler, 'Social Neurosis', p.75.
68. FLD16/147/81/2, Outrages, Deputation of Delegates from Outlying Districts, Pretoria, 6 September 1905, Report of Proceedings, Deputation to his Excellency the Lieutenant-Governor on the Question of Desertion of Chinese Labourers from the Mine. See similar complaints in FLD16/147/81/7, Outrages, Complaints by Ellewee, Davidson, Schoeman and Others to Botha, 13 January 1906, response from SAC, 8 February 1906.
69. FLD16/147/81/13, Lieutenant-Governor to Colonel Steele, 17 April 1906. The text in italics was originally underlined in the letter.
70. FLD16/147/81/13, Solomon to Malcolm, 19 February 1906.
71. FLD16/147/81/9, Tennant to Acting Lieutenant-Governor, 18 April 1906.
72. FLD16/147/81/13, Selborne to Solomon, 12 July 1906; reply, 13 July 1906.
73. FLD Dep. 56, Correspondence with various colonial governors and their secretaries, Fol.165, Personal, Sol to Sel, 11 April 1906.
74. FLD16/147/81/4, Resident Magistrate, Pretoria, to Lieutenant-Governor, 23 September 1905.
75. FLD24/26/6/1, Charles Moss to Elgin, 20 December 1905, Enclosed letter, Mag to Fannie, 18th October 1905.
76. FLD24/26/6/1, Charles Moss to Elgin, 20 December 1905, Enclosed letter, Mag to Fannie, 18th October 1905.
77. FLD24/26/6/1, Charles to Elgin, 27 December 1905.
78. Balfour-Browne, *South Africa* (1905), pp.148-149.
79. *Pretoria News*, 4 December 1906, p.4.
80. FLD10/147/36/8, E. M. Showers, Commissioner of Police, to Acting Secretary, Law Department, 3 April 1905.
81. See FLD135/23, Selborne to Lyttelton, 30 January 1905.
82. L. V. Praagh (ed.), *The Transvaal and its Mines* (London, 1906), p.537; FLD50/6/1, Report by Superintendent, FLD, on Disturbances at Aurora West, Geduld & Glen Deep, 25 October 1904; Report of the Disturbance at the Aurora West United Gold Mining Company Limited, 22 October 1904; Report on the Disturbance at Geduld Proprietary Mines Limited, 16 October 1904; FLD Report on the Disturbance at the Glen Deep Ltd, 19 October 1904.
83. FLD50/6/1, Verbatim Report, evidence given at enquiry held at Cason G. M. Co. by Chief Inspector Purdon, and Inspector Wilkinson, FLD, (formerly Acting Consul, Wuchow), 13-21 August 1907, into riot, 5 August 1907, Richard Charles Boyd statement.
84. FLD50/6/1, Mark Foxcroft statement.
85. FLD50/6/1, Richard Charles Boyd statement.
86. FLD50/6/1, William Herbert George Jackson, Controller at the Angelo Gold Mine, statement.

226 Notes

87. FLD50/6/1, James Morris statement.
88. FLD10/147/36/1, Superintendent's Report, Riots and Disturbances amongst Chinese Labourers, 23 November 1904 to 25 March 1905, pp.4–5.
89. *The Daily News*, 9 January 1906, by H. W. Massingham.
90. FLD50/6/4, Baldwin to Bagot, CMLIA, 30 September 1904.
91. FLD50/6/4, Superintendent, FLD, to Chairman, Randfontein Estates G. M. Co. Ltd., 19 September 1904.
92. FLD7/147/20/18, Jamieson to Malcolm, 18 September 1906.
93. FLD19/16/05, H. A. Young, Assistant Resident Magistrate 'B' Court, Johannesburg, to Secretary, Law Department, 15 August 1905.
94. FLD10/73/32, petition of #690,841,924,693,923,832,933,850 (Mine Nos.) of the New Heriot G. M. Co., on behalf of the labourers of that Mine, to Superintendent, FLD, translated by W. Zehnder, 9 April 1907.
95. FLD51/6/8, J. W. H. Stubbs, General Manager, Randfontein Estates Gold Mining Co. Ltd, to J. W. S. Langerman, Chairman, Randfontein Estates Gold Mining Co. Ltd, 19 October 1904.
96. FLD19/20/05, H. Tennant, Secretary, Law Department, to Superintendent, FLD, 13 October 1905.
97. Johannesburg, *Star*, 22 December 1905.
98. For examples, see HE250/139, C. W. Spence, General Manager, Consol. Main Reef Mines & Estates Ltd, to G. A. Goodwin, General Manager, WNLA, 2 April 1902; UKPP, Cd1950, Acting Commissioner Pearce, Zomba, to Lansdowne, 3 August 1903, p.327; Warwick, pp.172–174.
99. FLD10/147/36/8, Report into the disturbances at the North Randfontein Gold Compound, 17 September 1904.
100. HE251/138, Letter extract, 22 January 1904, forwarded by Perry, from Warnsford Loch, General Manager, The Raub Australian Gold Mining Co., Ltd., Pahang, Federated Malay States.
101. HE185/285, Schumacher to Wernher, Beit & Co., 6 June 1904.
102. Phillips, *Reminiscences*, p.112.
103. *East Rand Express*, 28 December 1907, p.29; 4 January 1908, p.30; *The Prince*, 5 January 1906, p.15. Many of the photographs and postcards of the scheme show such activities.
104. *Bloemfontein Post*, 21 September 1904. My italics.
105. FLD51/6/13, Superintendent to Private Secretary, Acting Lieutenant-Governor, 14 December 1904.
106. FLD51/6/13, E. M. Showers, Commissioner of Police, to Secretary, Law Department, 3 January 1904.
107. FLD51/6/17, Grant-Smith, draft report, June 1905. Unusually, no final report included. Other examples in FLD 10/147/36/8, Report into disturbances, North Randfontein Gold Compound, Saturday, 17 September 1904; FLD10/147/36/11, Jamieson, FLD, to Private Secretary, Lieutenant-Governor, 16 May 1905; FLD10/147/36/14, FLD Report of the Disturbance on 1–2 July 1905, amongst Chinese Coolies employed on the Wits Deep Ltd. by Inspector Fairfax, 5 July 1905; FLD60/6/3, Disturbance at the Van Ryn Gold Mines & Estate Ltd., FLD report from Grant-Smith, Inspector, 1905.
108. FLD10/147/36/8, Disturbances and Riots, report into the disturbances at the North Randfontein Gold Compound, 17 September 1904.

109. FLD147/20/13, Allegations by Johnstone of ill-treatment at New Kleinfontein, Compound Magistrates Report and Evidence, 24–26 March 1906, Alfred Child testimony.
110. See Chapter 3.
111. See the glowing description by the touring British Association of Science in South Africa, written Sir William H. Preece, *Society of Arts Journal*, 54 (17 November 1905–16 November 1906), p.52.
112. Trew, pp.103–104; see also Polley, p.85.
113. FLD9/147/36/19, Secretary, Lieutenant-Governor, to Secretary, Chamber of Mines, 11 May 1905.
114. FLD9/147/36/19, F. Raleigh, Acting General Manager, Rand Mines Ltd to Manager, Rose Deep Mine, 10 May 1905.
115. FLD10/147/36/10, Acting Secretary, Native Affairs, to Lieutenant-Governor, 31 March 1905.
116. FLD10/147/36/10, H. F. Petersen, General Manager, Geldenhuis Estate, to District Controller, Native Affairs Department, Germiston, 27 March 1905.
117. FLD9/147/36/8, Report into disturbances at the North Randfontein Gold Compound, 17 September 1904.
118. See E. Webster, (ed.), *Essays in Southern African Labour History* (Johannesburg, 1978), p.12.
119. FLD15/147/56, Acting Secretary, Native Affairs, to Secretary, Lieutenant-Governor, 17 January 1905; see also N. Devitt, *Memoirs of a Magistrate* (London, 1934), p.70.
120. FLD51/6/13, J. A. Boyd, Secretary, Wits Gold Mining Co. Ltd (Knight's), to Solomon, Acting Lieutenant-Governor, 5 January 1905.
121. FLD51/6/16, Manager, Nourse Mines Ltd., to Inspector, FLD, 25 September 1908.
122. For reports of such attacks, see UKPP, Cd1897, Labour Commission Evidence, p.608; *The Transvaal Advertiser Pretoria*, 18 July 1906.
123. Chamber of Mines Annual Report, 1905, p.50.
124. FLD7/28/6/05, Wolfe-Murray, Report on the case of labourer 27992 (Hai Sung Ling) Durban Roodepoort Deep Ltd., undated.
125. FLD7/28/6/05, Secretary, FLD, to Lieutenant-Governor, 28 March 1905.
126. FLD7/28/6/05, Evans, Superintendent, FLD, to Hutton, Manager, Van Ryn Mine, 14 March 1905.
127. FLD7/147/20/11A, Lawley to Chaplin, 13 June 1905.
128. *British Friend* 1905, quoted in Hope Hay Hewison, *Hedge of Wild Almonds, South Africa, The Pro-Boers & the Quaker Conscience, 1890–1910* (London, 1989), p.275.
129. FLD50/6/3, Jamieson to Solomon, 30 September 1905.
130. FLD51/6/15, General Manager, Kleinfontein Group Central Administration, Benoni, to Jamieson, 25 August 1905.
131. FLD51/6/10, Lane Carter, Acting Manager, French Rand G. M. Co. Ltd, to C. J. Price, General Manager, Central Administration, Johannesburg, 10 April 1905.
132. FLD125/16/25, Arrival Reports, S/S Katherine Park, 13 October 1905.
133. Maryna Fraser (ed.), *Johannesburg Pioneer Journals, 1888–1909*, (1985), Bright to Mother, 10 September 1905, pp.184–185. Assistant to the head surveyor at Vogelstruis Consolidated Deep, Ltd.

134. Breckenridge, p.674.
135. FLD8/147/29, FLD Coolies wives and children, 1904–06.
136. *Pretoria News*, 12 November 1906, p.5.
137. James Rose Innes, *Autobiography* (London, 1949), p.218. Karen Harris, 'Private and Confidential: The Chinese Mine Labourers and "Unnatural Crime"', *South African Historical Journal*, L (2004), pp.124–128.
138. Richardson, *Chinese*, p.127; Louis F. Free, *The Problem of European Prostitution in Johannesburg, A Sociological Survey* (Johannesburg, 1949), p.9. Richardson's source was the Chamber of Mines records, which have since been destroyed by the company.
139. Harries, *Culture*, p.201 & 'Symbols', p. 326.
140. Grant wrongly asserts that the CO did not care, p.97, ft.75.
141. FLD15/147/56, Evans to Secretary, Acting Lieutenant-Governor, 7 January 1905.
142. FLD15/147/56, Acting Secretary, Native Affairs, to Secretary, Lieutenant-Governor, 17 January 1905.
143. See the folder, SNA248/70/05, Contact of Chinese with Kaffir Women.
144. FLD179/36/36.
145. *Transvaal Critic*, January 1906 in particular.
146. FLD179/36/36, Handwritten note, signed 'G. E.', 8 February 1907, my italics.
147. FLD25/86/06, C. D. Stewart, Inspector Interpreter, to Jamieson, 22 January 1906; Selborne to Elgin, March 1906.
148. FLD16/147/81/16, Police Report, F. L. Gaum, 11 July 1906 re Roos' family; Report, District Surgeon, H. J. Leonard, Roodepoort, 11 July 1906.
149. FLD16/147/81/16, Outrage on Roos Family, Roodpoort, 11 July 1906, Copy, Report from District Surgeon, H. J. Leonard, Roodepoort.
150. FLD16/147/81/16, Commissioner of Police, Transvaal Town Police, to Malcolm, 13 July 1906.
151. McCulloch, p.28.
152. McCulloch, p.4.
153. FLD16/147/81/16, Outrages, Outrage on Roos Family, Roodpoort, 11 July 1906, Copy, Report from District Surgeon, H J Leonard, Roodepoort, 13 July 1906, Commissioner of Police, Transvaal Town Police, to Malcolm.
154. FLD25/86/06, Chief Magistrate, Johannesburg, to Secretary, Law Department, 5 January 1906.
155. See photographs in Scully, *Ridges of White Waters*, pp.215, 219; *The Transvaal Critic*, 15 June 1906.
156. CO537/540, pp.49–51, Report of an Enquiry held by Mr J. A. S. Bucknill into certain allegations as to the prevalence of unnatural vice and other immorality amongst the Chinese indentured labourers employed on the mines of the Witwatersrand. 20 September 1906.
157. Saloni Mathur, 'Wanted Native Views: Collected Colonial Postcards of India', in A. Burton (ed.), *Gender, Sexuality, and Colonial Modernities* (1999), p.96.
158. McCulloch, p.4.
159. CO537/540, 13 August 1906, p.47; W. F. Trew, Inspector, District Commandant, Pretoria, to R. S. Curtis, SAC, p.28; see also 19 August 1906, E. M. Showers to R. Solomon, p.45.

160. Anderson describes a similar investigation in Kenya, p.57.
161. CO537/540, Bucknill Report, N. Audley Ross to Malcolm, 18 August 1906.
162. CO537/540, Bucknill Report, p.265.
163. CO537/540, Bucknill Report, p.268.
164. CO537/540, Bucknill Report, pp.270–271.
165. CO537/540, Bucknill Report, pp.12–13.
166. CO537/540, Bucknill Report, pp.14, 17.
167. CO537/540, Bucknill Report, C. B. Furse to Malcolm, 8 August 1906, p.22.
168. CO537/540, Bucknill Report, E. T. Campbell, Acting Mayor, Roodepoort-Maraisburg Municipality, 13 August 1906, p.26; see also Harries, *Culture* and Foreman, 'Randy on the Rand'.
169. CO537/540, Bucknill Report, C. Brammer, Germiston Municipality, Town Clerk's Office, 15 August 1906, p.28.
170. CO537/540/35, 24 August 1906, N Audley Ross to Selborne.
171. CO537/540, Bucknill Report, p.67.
172. CO537/540, Bucknill Report, p.68.
173. CO537/540, Bucknill Report, 20 September 1906, p.70.
174. CO537/540/38767, Selborne to Governor, 1 October 1906.
175. CO537/540, Bucknill Report, R. McNally, Deputy Mayor, Krugersdorp, Town Clerk's Office, to Malcolm, 15 August 1906, p.29. See also Harris, pp.128, 130.
176. CO537/540, Bucknill Report, Report, pp.5–6.
177. CO537/540, Bucknill Report, amended FLD Memorandum, forwarded to Malcolm on 21 August 1906, pp.53–55.
178. FLD168/34/1, Excerpt from a letter, 23 April 1906, R. Solomon to Supt., FLD
179. Forman, Ross, G., 'Randy on the Rand: Portuguese African Labor and the Discourse on "Unnatural Vice" in the Transvaal in the Early Twentieth Century', *Journal of the History of Sexuality*, 11:4 (2002), 570–609.
180. Forman, p.581.
181. The terms 'natives' and 'aboriginals and coloureds' were often used interchangeably by the judiciary and the FLD; it is unclear in these cases, however, whether these are indeed the same groups or whether one set of statistics refers only to 'natives' and the other combines Cape Coloureds and 'natives'.
182. FLD14/147/52/1–8, Convictions and Sentences; LD1301/3183/06, Convictions of Chinese Labourers and Aboriginal Natives, 1 July 1904–30 June 1905; PM16/24/5/1907, Chinese, Convictions and Sentences; Chamber of Mines *Annual Report*, 1905, pp.xxxii–xxxiii, 51.
183. FLD14/147/52/1–8, Convictions and Sentences; FLD16/147/81/8, Murders committed, 1 January–30 June 1906; LD1301/3183/06, Convictions of Chinese Labourers and Aboriginal Natives, 1 July 1905–30 June 1906; PM16/24/5/1907, Chinese, Convictions and Sentences; Chamber of Mines *Annual Report*, 1905, pp.xxxii–xxxiii, 51.
184. FLD14/147/52, Convictions and Sentences on Chinese Labourers from 1st July 1905 to 30th June 1906.
185. FLD16/147/81/11, Tennant to Lieutenant-Governor, 11 May 1906.
186. Fraser (ed.), 'Edmund Bright Letters', *Bright to Mother*, 10 September 1905, pp.184–185.

230 Notes

187. Hermann Giliomee, 'The Beginning of Afrikaner Nationalism, 1870–1915', *South African Historical Journal*, 19 (1987), p.124.
188. While no formal African complaints were made, several Afrikaner farmers cited brutal attacks upon their African tenants or neighbours. See FLD24/26/6/1, Mag to Fannie, 18 October 1905; FLD10/147/36/8, E. M. Showers, Commissioner of Police, to Acting Secretary, Law Department, 3 April 1905; FLD16/147/81/2, Deputation of Delegates from Outlying Districts, Wolmarans' statement, 6 September 1905.
189. Johannesburg *Star*, 5 April 1905.
190. Innes, p.218.
191. *The Times*, 4 November 1908.
192. see Chapter 7.
193. FLD4/147/9, R. S. Curtis, Inspector General, SAC, to Lieutenant-Governor, 19 April 1906.
194. *East Rand Express*, 7 September 1907, p.13, says 'Chinese Outrages. WELL-KNOWN RESIDENT KILLED.'
195. See FLD49/5/93, Repatriation of Chinese Labourers, Last Batch, SS Heliopolis, 8 March 1910 departure: Secretary, Law Department, to Jamieson, 10 February 1906.
196. McCulloch, p.4.
197. Cornwell, pp.441–442.
198. See Frantz Fanon, *The Wretched of the Earth* (1963).

6 Adapting the Stereotype: Race and Administrative Control

1. For more on archives in South Africa, see Carolyn Hamilton et al. (eds.), *Reconfiguring the Archive* (2002).
2. Jon E. Wilson, 'Agency, Narrative, and Resistance', in Sara Stockwell (ed.), *The British Empire*, p.256.
3. Wilson, p.255.
4. Frederick Cooper, 'Conflict and Connection: Rethinking Colonial African History', *The American Historical Review*, 99:5 (December 1994), pp.1516–1545.
5. HE250/140, Skinner to WNLA, 13 June 1903.
6. F. Cooper & A. L. Stoler, 'Between Metropole and Colony: Rethinking a Research Agenda', in Cooper & Stoler, p.8.
7. F. Cooper & A. L. Stoler, 'Between Metropole and Colony', p.8.
8. MP10/5/85, Milner to Lyttelton, 23 January 1905.
9. Innes, p.218.
10. See Chapters 4 and 7.
11. Phillips, *Transvaal Problems*, pp.106–108.
12. See Chapter 5 for numerous examples of this.
13. FLD225/62/55, A. S. H. Cooper to Selborne regarding allegations by Mr Cooper whilst in Gaol, 8 February 1906; FLD50/6/1, 3 March 1909, Acting Supt., R I Purdon, to J de Villiers Roos Esq., Acting Secretary to the Law Department.
14. *The Mercury* (Tasmania), 24 October 1906, p.5; 25 May 1905, p.5.
15. FLD50/6/3, Report into Disturbance on 2 December by Evans, 13 December 1904.

16. FLD50/6/3, Evans to General Manager, General Mining & Finance Corp. Ltd, 13 December 1904.
17. FLD50/6/3, Evans to General Manager, General Mining & Finance Corp. Ltd, 20 December 1904.
18. FLD50/6/3, Manager of Mine, General Mining & Finance to Evans, 19 December 1904.
19. Devitt, pp.70–71.
20. Robert Kennaway Douglas, 'China', in *Encyclopaedia Britannica* (1902), quoted in Mackerras, pp.69–71.
21. FLD50/6/1, William Herbert George Jackson, Controller at the Angelo Gold Mine, statement; see also FLD50/6/1, Enquiry at the Cason G. M. Co., Mark Foxcroft statement.
22. HE251/138, Circular: Extract from letter from Hamilton, 20 April 1904; see also HE251/138, letter from Perry and Hamilton, 26 February 1904.
23. David Ownby, 'Recent Chinese Scholarship on the History of Chinese Secret Societies', *Late Imperial China*, 22:1 (June 2001), pp.139–158.
24. FLD12/147/41/1, Report on the 'Ko Lo Hai' by E. Lenox Simpson, Chinese Controller, French Rand Gold Mining Company, Ltd.; Evans to Acting Lieutenant-Governor, 6 February 1905; FLD10/147/36/1, Riots and Disturbances amongst Chinese Labourers, Report, Superintendent, FLD, 23 November 1904–29 March 1905, p.5; FLD4/147/8, Formation of Secret Societies by Chinese.
25. FLD10/147/36/1, Riots and Disturbances amongst Chinese Labourers, Report, Superintendent, FLD, 23 November 1904–29 March 1905, p.4.
26. FLD50/6/1, Enquiry at the Cason G. M. Co., James Morris, Manager, ERPM, statement.
27. Chilvers, pp.167–168.
28. FLD50/6/1, Inspector of Police, Boksburg, to Superintendent, FLD, 23 October 1906; Governor to Colonial Secretary, 18 May 1906; FLD6/147/18, Mortality on Mines.
29. FLD10/147/36/8, George Wolfe Murray, Acting Superintendent, FLD, to Private Secretary, Lieutenant-Governor, 5 April 1905.
30. FLD20/52/05, Jamieson to Secretary, High Commissioner, and Solomon, 19 August 1905.
31. Devitt, p.70.
32. HE221/51, Bagot, CMLIA, Coolie Absences Circular, 2 August 1907.
33. FLD9/147/30/1A, Report of Senior Medical Supt. and Stuart Knaggs, Senior Superintendent, SS Sikh, 16 December 1904.
34. Skinner Report, p.81.
35. FLD7/147/20/15, Private Secretary, FLD, to Lieutenant-Governor, 30 May 1906; See Chapter 5 for a further breakdown of the role of the Chinese police.
36. FLD25/65/06, Case of Li Koui Yu vs. Supt. FLD, Solomon to Jamieson, 23 March 1905.
37. FLD168/34/1 (2nd), Solomon to Jamieson, 17 April 1906.
38. FLD25/65/06, Case of Li Koui Yu vs. Supt. FLD, Solomon to Jamieson, 23 March 1905.
39. FLD168/34/1 (2nd), Solomon to Jamieson, 17 April 1906; FLD125/16/22, Surgeon-Superintendent's Report SS Lothian, Edmond Lawless, 30 May 1905; FLD125/16/22, Arrival Reports, Bagot to Jamieson, 2 June 1905.

40. FLD8/147/21, Jamieson to Solomon, 2 July 1906.
41. FLD8/147/21, Private Secretary, Lieutenant-Governor's Office, to Jamieson, 23 July 1906.
42. CO537/54035, N. Audley Ross to Selborne, 24 August 1906, p.36.
43. FLD19/20/05; FLD1/147/1/1–3.
44. FLD 1/147/2, GGR [Robinson], Governor's Office, to 'My dear Geoffrey' [Lagden], 2 June 1904; See also Hanretté, Secretary, Law Department, to Secretary, Lieutenant-Governor, 21 May 1904; FLD179/36/39, Captain R. R. Gibson, Inspector, FLD, to Superintendent, FLD, 16 September 1907.
45. *The Transvaal Advertiser Pretoria*, 31 July 1903, Legislative Council Debate.
46. FLD12516/22, Arrival Report, SS Lothian, 30 May 1905.
47. FLD122/16/6, Arrival Report, SS Tweedale, 7 October 1904; FLD9/147/30/1A, Health Report, S/S Courtfield, 30 November 1904.
48. FLD9/147/30/1B, Health Report, Hoggan, Surgeon Superintendent, S/S Sofala, 28 December 1904.
49. *Rand Daily Mail*, 13 November 1906.
50. FLD9/147/30/1A, Milner to Lyttleton, 19 December 1904.
51. FLD1/147/2, Hanretté, Acting Secretary, Law Department, to Lieutenant-Governor, 9 March 1905.
52. See the glowing description by the touring British Association of Science in South Africa, written Sir William H. Preece, *Society of Arts Journal*, 54 (17 November 1905–16 November 1906), p.52.
53. Trew, pp.103–104; see also A. E. Polley, *John Chinaman on the Rand. A New Form of Torture* (London, 1905), p.85.
54. FLD51/6/13, J. A. Boyd, Secretary, Wits Gold Mining Co Ltd (Knight's), to Solomon, Acting Lieutenant-Governor, 5 January 1905.
55. FLD51/6/16, Manager, Nourse Mines Ltd., to Inspector, FLD, 25 September 1908.
56. SP62/10, Selborne to Farrar, undated, 1906.
57. SP62/157–8, Selborne to Phillips, 15 January 1906.
58. SP62/159–64, Memo on Position of White Miners, undated.
59. Reprinted in *The Sydney Morning Herald* (NSW), 1 December 1905, p.7.
60. HE221/51, Bagot, Memorandum for the Executive Committee of the Chamber of Mines, 22 March 1906.
61. HE221/51, General Manager's Report, to Board of Management, CMLIA, 20 December 1905.
62. van Onselen, vol. II, *New Nineveh*, p.31.
63. SP62/185–92, Phillips to Selborne, 24 January 1906.
64. Kubicek, *Economic*, pp.129, 137–139.
65. SP62/12, Farrar to Selborne, 4 May 1906.
66. HE221/50, Legislative Council Fifth Session, 1905, Labour Importation Draft Ordinance; *The Transvaal Advertiser Pretoria*, 26 July 1906, 2nd reading of Labour Importation Amendment Draft Ordinance.
67. HE221/50, CMLIA, G. L. Craik, Notes on proposed Amendments to the Labour Importation Ordinance 1904, September 1905.
68. Trew, p.103.
69. FLD168/34/1(2nd), Jamieson to de Jong, President, Chamber of Mines, 28 February 1906.

70. MP192, Selborne to Milner, 7 May 1906; SP62/14, Selborne to Farrar, 7 May 1906.
71. Yudelman, p.264.
72. FLD50/6/1, Private Secretary to Secretary of the Association of Mines Managers, Transvaal, undated [1905], p.2.
73. HE221/50, H. Eckstein & Co. to Superintendent, FLD, 27 November 1905.
74. HE89/222. H. Eckstein to Wernher, Beit & Co., 10 July 1905.
75. HE134/64, Report of meeting, Evans to Phillips, 28 August 1905.
76. HE221/57, Walter Bagot to Board of Management, CMLIA: General Manager's Report, 5 December 1905.
77. FLD24/27/06/2, Bagot, CMLIA, to Jamieson, 21 February 1906.
78. FLD4/147/9, Secretary, Lieutenant-Governor, to Secretary, Chamber of Mines, 10 November 1904.
79. FLD4/147/9, Secretary, Chamber of Mines, to Secretary, Lieutenant-Governor, 6 January 1905. See also FLD4/147/9, Secretary, Chamber of Mines, to Lieutenant-Governor, 3 November 1904, pp.1–2, for a similar disagreement.
80. HE221/51, Jamieson to General Manager, CMLIA, 21 April 1906; HE221/51, General Manager, Rand Mine ltd, memo for L. Reyersbach, re Chinese Enquiry Houses on Mines.
81. FLD4/147/9, Telegram, Selborne to Lyttelton, 1 April 1905.
82. HE221/51, Bagot to Board of Management and Mines Employing Chinese, 24 January 1906.
83. FLD19/37/05, Memorandum as to future supply of Inspectors for the FLD and of TEAs in China, 30 November 1905, pp.3–4.
84. FLD125/16/22, Bagot to Superintendent, FLD, 2 June 1905.
85. FLD4/147/7/1, Lyttleton to Selborne, 7 September 1905.
86. FLD17/147/93/B, Government Supervision of Labourers: Cowie, Secretary, Chamber of Mines, to R. Solomon, 26 June 1906.
87. FLD17/147/93/A, Cowie, Secretary, Chamber of Mines, to Solomon, 17 May 1906.
88. FLD17/147/93/A, Solomon's response, 19 May 1906.
89. FLD17/147/93/B, Cowie, Secretary, Chamber of Mines, to R. Solomon, 26 June 1906.
90. FLD17/147/93/A, Secretary, Lieutenant-Governor, to Secretary, Chamber of Mines, 7 July 1906.
91. Wilson, p.256.

7 Political Repercussions: Self-Government Revisited

1. Balfour-Browne, p.47.
2. CO179/189, quoted in Lake & Reynolds, p.122. See Chapter 1 about the test.
3. *The Spectator*, 22/x/1904, letter from Peter Green, quoted in McCallum, p.51.
4. Russell, p.78 and Figures 2 and 3, p.79.
5. Thompson, *Empire*, pp.69–71; Katz, *Trade*, p.111.
6. Russell, pp.107, 137–143.
7. *The Advertiser*, 10 January 1906, p.6.
8. *The Brisbane Courier*, 26 February 1906, p.4.

9. Earl Grey, quoted in Ronald Hyam, 'Smuts and the Decision of the Liberal Government to Grant Responsible Government to the Transvaal, January and February 1906', *The Historical Journal*, 8:3 (1965), pp.385, 387.
10. Cd2788, Selborne to Elgin, 20 December 1905, p.2.
11. Marlowe, pp.166–167.
12. Cd2788, Elgin to Selborne, 20 December 1905, p.2.
13. HE90/72-4, Phillips to Wernher, undated; MP192/3–7, Selborne to Milner, 7 May 1906. While both Jamieson and Selborne expected only 200–250 Chinese to sign up, the mine managers and compound managers thought most would choose to return to China.
14. See Chamber of Mines *Annual Reports*, 1905, 1906, 1907; Phillips, *Some Reminiscences*, p.138.
15. HE202/12, Wernher, Beit & Co to Eckstein & Co, Private, 17 May 1906.
16. FLD12/147/41/2B.
17. FLD12/147/41/2A, May 1906: Jumpers Deep, French Rand, Rose Deep, Van Dyk Proprietary Mines (Boksburg), Glen Deep, Ginsberg Gold Mining Co., the Consolidated Main Reef Mines, and Wits. Gold Mining Co.
18. FLD12/147/41/2A, May 1906.
19. FLD12/147/41/2B.
20. Fraser (ed.), Bright to Mother, 2 July 1906, p.186.
21. FLD12/147/41/2A.
22. SP163/62/6–7, Farrar to Selborne, 2 April 1906.
23. FLD12/147/41/2A.
24. Davenport, *Afrikaner*, pp.257–258.
25. Reprinted in *The Register* (Adelaide), 28 March 1905, p.3.
26. See coverage in *The Sydney Morning Herald*, 25 October 1905, p.9.
27. *The Sydney Morning Herald*, 22 January 1906, p.6.
28. Reprinted in the *Wanganui Chronicle*, 22 January 1906, p.7.
29. Jebb, pp.131–132.
30. B. K. Long, *Drummond Chaplin* (London, 1941), p.86. See Chapter 3.
31. S. C. Cronwright-Schreiner (ed.), *The Letters of Olive Schreier, 1876–1970* (London, 1924), Schreiner to Mrs. Francis Smith, p.247.
32. *Transvaal Advertiser*, 17 August 1906; Mawby, *Gold II*, pp.479–480.
33. Mawby, *Gold II*, pp.714–715; *Star*, 6 March 1906.
34. *Poverty Bay Herald*, 25 April 1906, p.4.
35. *Auckland Star*, 9 February 1906, p.3. Seddon was prevented from taking further due to dying in June 1906; *Evening Post*, 4 May 1905, p.4.
36. *Marlborough Express*, 18 January 1906, p.2.
37. *Tuapeka Times*, 21 February 1906, p.3.
38. *Auckland Star*, 10 March 1906, p.10; *The Sydney Morning Herald*, 22 September 1906, p.17; see also 14 September 1906, p.3.
39. FLD12/147/41/2B, Jamieson to R. Solomon, 12 July 1906.
40. Richardson, *Chinese*, p.183; *Transvaal Advertiser*, 6 August 1906.
41. *Star*, 25 November 1903–20 April 1905.
42. Mawby, *Gold II*, pp.484–492, 540–541; MP10/19–21, Milner to Lyttelton, 9 May 1904; MP10/44, Milner to Lyttelton, 25 July 1904.
43. Mawby, *Gold II*, p.614.
44. Transvaal Responsible Government Association, *Important Declaration of Policy, Common Action with 'Het Volk' to Obtain Responsible Government, Manifesto* (1905); Mawby, *Gold II*, pp.584–588.

45. *Leader*, 18 June 1904.
46. *Rand Daily Mail*, 22 November 1904; *Leader*, 30 November 1904.
47. *Star*, 20, 24 October 1906.
48. R. Churchill, Selborne to Churchill, 2 March 1906, p.510.
49. *Star*, 25 April 1905, p.6.
50. SP163/60/4, Memorandum on the effect of Arresting the Importation of Chinese Labour; *Transvaal Leader*, 15 January 1906; Long, p.86.
51. *Star*, 20 October 1906.
52. MP191/157-167, Perry to Milner, 5 March 1906; R. Churchill, Abe Bailey to Churchill, 8 January 1906, p.501.
53. Mawby, 'Capital', 398; Duminy & Guest, 'Progressive Pioneer' to Editor, *Star*, 26 March 1906, pp.426-427.
54. *The Prince*, 5 January 1906, p.3.
55. *Transvaal Advertiser*, 14 August 1906, letter from A. Brittlebank; see also Mawby, *Gold II*, pp. 542-550, 666-667, 671-672; Jeeves, pp.10-11; HE338/7, Eckstein to Phillips, 23 November 1906.
56. Long, pp.587-589; G. G. Robinson, 'Political Parties in the Transvaal', *National Review*, 45 (1905), p.474.
57. *Star*, 27 October 1906, Reform Club meeting.
58. See the tours of Raymond Schumacher, who quit as director at Wernher, Beit & Co. to campaign for Chinese labour, and Creswell in 1905-06. Both spoke at the London Liberal Club on 21 February 1906 and their letters to *The Times* and the *Westminster Gazette*, January 1906; See Raymond W. Schumacher, *A Transvaal View of the Chinese Labour Question* (London, 1906); F. H. P. Creswell, *The Witwatersrand Gold Mines and Responsible Government: Cause of the Depression* (Cape Town, 1906); Robert Raine, (ISAA), *Transvaal Labour: Unskilled whites: Manager of Village Main Reef on Mr. Creswell: Mr. Creswell's Statements Critically Examined and Refuted* (London, 1906).
59. *Transvaal Advertiser*, 5 July 1906.
60. *Transvaal Advertiser*, 7 July 1906.
61. *Star*, 15, 16, 26 September 1905.
62. *Transvaal Leader*, 28 January 1905. See Davenport, *Afrikaner*, p.256.
63. Garson. pp.222, 224.
64. D. W. Kruger, *The Making of a Nation: A History of the Union of South Africa, 1910-1961* (1969), p.21.
65. Jeeves, p.13.
66. Giliomee, *Afrikaners*, p.244.
67. Hancock, van der Poel & Smuts to M. C. Gillett, 25 May 1906, p.277.
68. Walker & Weinbren, p.16.
69. *Bloemfontein Post*, 7 November 1904 and thereafter.
70. FLD50/6/1, 12 and 18 April 1906.
71. FLD50/6/1, Pretoria Meeting, 4 May 1906.
72. *Star*, 1 June 1905.
73. Cd2823, Letter of Instructions to the Chairman of the Committee appointed to Enquire and Report upon certain matters connected with the future constitutions of the Transvaal and ORC, pp.3-4; Hyam & Marten, *Reappraisals*, pp.139-140, David E. Torrance, *The Strange Death of the Liberal Empire: Lord Selborne in South Africa* (1996), pp.74-77; W. S. Churchill, *A Note upon the Transvaal Constitution Question*, 30 January 1906.

74. Hancock, van der Poel & de Villiers to Smuts, 25 June 1906, pp.288–289.
75. George D. Boyce, (ed.), *The Crisis of British Power: The Imperial and Naval Papers of the Second Earl of Selborne, 1895–1910* (1990), Selborne to Elgin, 28 March 1906, p.258.
76. Mawby, *Gold II*, p.717.
77. Duminy, Guest & Fitzpatrick to Phillips, 4 December 1904, pp.380–381.
78. *Transvaal Advertiser*, 27, 28 August 1906.
79. Transvaal Report of the Committee re Pretoria Indigents 1905, in Mawby, *Gold II*, pp.699–700.
80. Garson, p.112.
81. *Transvaal Advertiser*, 5 July 1906.
82. *Leader*, 20 April 1905; Mawby, *Gold II*, p.617.
83. *Transvaal Advertiser*, 5 July 1906.
84. Mawby, *Gold II*, p.732.
85. *Star*, December 1905, Letter to Editor, J. W. Quinn, Mawby, II, pp.618–619.
86. Mawby, *Gold II*, pp.737–738.
87. Burt, pp.73–75; Mawby, *Gold II*, pp.570–573.
88. *Star*, 31 May, 14–15 September 1905; Mawby, *Gold II*, p.630.
89. Election results from Cd3528, pp.162–166; Ernest Clough, *The South African Parliamentary Manual* (London, 1909); O. F. Brothers, *The First Transvaal Parliament* (Johannesburg, 1907).
90. Solomon, Edward Rooth to Molteno, 18 March 1907, p.285.
91. MP226/138-54, Private Memorandum by Fitzpatrick on Chinese labour, 25 March 1907.
92. Jeeves, A., Het Volk and the gold mines: the debate on labour policy, 1905–1910, University of the Witswatersrand, African studies seminar paper, 92 (1980), pp.32–33; Wheatcroft, p.223.
93. Long, pp.106–107.
94. MP226/138-54, Private Memorandum by Fitzpatrick on Chinese labour, 25 March 1907.
95. Yudelman, pp.261–262.
96. Jeeves, Het Volk, p.33.
97. PM52/3/1907, Acting Secretary, Prime Minister, to District Medical Officer of Health, Johannesburg, 18 April 1907; PM24/6/1907, Acting Secretary, Prime Minister, to Secretary, Law Department, 5 April 1907; Appendix B.
98. Beinart; Breckenridge.
99. Carl Jeppe, *The Kaleidoscopic Transvaal* (London, 1906), p.250; see also F. Eckstein to L. Phillips, 3 October 1913, quoted in Kubicek, *Economic*, p.80.
100. *Star*, 17 March 1905, Company meeting of shareholders for South Geldenhuis Deep Consolidated Gold Fields Chaplin presided.
101. *Star*, 22 March 1905, p.9.
102. W. Basil Worsfold, *The Reconstruction of the New Colonies under Lord Milner*, volume I (London, 1913), p.367.
103. SP57/210-221, Selborne to Duncan, 8 July 1907.
104. T. Keegan, 'White Settlement and Black Subjugation on the Highveld', in William Beinart, Peter Delius, & Stanley Trapido (eds.), *Putting a Plough to the Ground, Accumulatopm and Dispossession in Rural South Africa, 1850–1930* (1986), p.244.
105. Chamber of Mines *Annual Report*, 1905, p.6; Long, p.111.

106. Pyrah, p.197 – these are approximations.
107. Simon Katzenellenbogen, *South African and Southern Mozambique: Labour, Railways and Trade in the Making of the Relationship* (1982), p.74.
108. *Star*, 1 May 1905, p.8.
109. HE221/51, Extract from Letter Received from Mr J E Brazier, 3 May, 1905.
110. Yudelman, p.258.
111. Lake, 'The White Man under Siege', pp.41–62; 'Translating', pp.199–219.
112. Jeeves, Het Volk, p.37.
113. Jeeves, Het Volk, p.12.
114. See Katz, *Labour*.
115. Yudelman, pp.265–267.
116. Jeeves, Het Volk, p.19.
117. *East Rand Express*, 14 September, 19 October 1907; 25 January 1908.
118. *East Rand Express*, 28 November 1907.
119. Webster, p.39.
120. *SAN*, 3 June 1905.
121. *SAN*, 3 June 1905; *Star*, 20 April 1905; Elaine N. Katz, 'White Workers' Grievances and the Industrial Colour Bar, 1902–1913', *South African Journal of Economics*, 42:2 (June 1974), p.148.
122. Kennedy, p.28.
123. *Transvaal Leader*, 27 May 1907; Mines Department *Annual Report, 1911*, Katz, 'White', pp.150–153.
124. McCallum, p.1; Katz, *Trade*, pp.147–149. His white supremacist, rather than socialist, ideology eventually lost him support within the party.
125. Giliomee, *Afrikaners*, pp.327–328, 331.
126. Jeremy Krikler, *White Rising: The 1922 Insurrection and Racial Killing in South Africa* (2005).
127. See, for instance, Hancock & van der Poel, Enclosure, R. Solomon to Smuts, 16 September [1908], p.321; Smuts to Merriman, 28 November 1906, p.304.
128. Davenport, *Afrikaner*, pp.271–272.
129. Lewsen, *Correspondence*, Merriman to Steyn, 10 March 1906, p.21.
130. Hancock, van der Poel & Smuts to Merriman, 30 August 1906, p.299.
131. Lewsen, *Correspondence*, Merriman to Steyn, 27 October 1905, p.9.
132. MmP26/63, Botha to Merriman, 25 June 1907.
133. MmP71/3/327, Merriman to Steyn, 12 February 1910.
134. MmP71/3/153, Merriman to Steyn, 20 April 1909.
135. MmP71/3/423, Merriman to Steyn, 22 May 1910.
136. Lewsen, *Correspondence*, Steyn to Merriman, 10 November 1905, p.10; MmP71/3/127, Merriman to Steyn, 18 March 1909.
137. Giliomee, *Afrikaners*, p.244.
138. MmP25/205, Steyn to Merriman, 10 November 1905.
139. *East Rand Express*, 21 September 1907, p.5.
140. Amery, *Political Life*, p.319.
141. Dubow, 'Imagining', pp.91–92. See also Saul Dubow, 'Colonial Nationalism, the Milner Kindergarten and the Rise of "South Africanism": 1902–1910', *History Workshop Journal*, 43 (1997), 57.
142. This was only meant to be temporary. David E. Torrance, 'Britain, South Africa, and the High Commission Territories: An Old Controversy Revisited', *The Historical Journal*, 41:3 (September 1998), pp.751–772;

Ronald Hyam, 'African Interests and the South Africa Act, 1908–1910', *Historical Journal*, 13:70 (1970), pp.85–105; Evans et al., p.175.
143. SP62/43–4, Farrar to Selborne, 27 October 1908.
144. MmP71/3/127, Merriman to Steyn, 18 March 1909.
145. Dobbie, p.797.
146. Edgar H. Brookes & Colin de B. Webb, *A History of Natal* (1965, 2nd ed., 1987), p.233; Worsfold, *Reconstruction*, p.383.
147. Harrison, pp.170–171.
148. Thompson, *Unification*, pp.82–94; Brookes & Webb, pp.234, 245; Dubow, 'Colonial', p.59.
149. Thompson, *Unification*, p.293.
150. See, for instance, Peter Merrington, 'Masques, Monuments and Masons: The 1910 Pageant of the Union of South Africa', *Theatre Journal*, 49:1 (1997), pp.3–4; Richard Solomon, 'The Resources and Problems of the Union of South Africa', *Royal Society of Arts Journal*, 59 (18 November 1910–10 November 1911), p.420.
151. Davenport, *Afrikaner*, p.299.
152. See the South Africa Office of Census and Statistics figures in Yap & Man, p.208. For instance, there were 2457 Chinese in the 1904 census but by 1921, this had dropped to 1828, and was still only 4340 in 1946. By the latter statistic, almost half were women and any woman married to a Chinese man was also recorded as being Chinese.
153. Devitt, pp.70–71.
154. Yap & Man, pp.135, 148–149.
155. Yap & Man, pp.103–104; Darryl Accone, *All Under Heaven, The Story of the Chinese Family in South Africa* (2004).
156. Dubow, 'Imagining', p.76.
157. Cell, p.57, 62.

Conclusion: Racialising Empire

1. *Indian Opinion*, 10 March 1906, p.142.
2. Yap & Man, p.95.
3. M. K. Gandhi, *An Autobiography*, p.236.
4. Duminy & Guest, Fitzpatrick to his wife, 28 July 1906, pp.478–479.
5. Hancock & van der Poel, Enclosure, R. Solomon to Smuts, 16 September [1908], p.406.
6. Karen Harris, 'Gandhi, the Chinese and Passive Resistance'; Harris & F. N. Pieke, 'Integration or Segregation: The Dutch and South African Chinese Compared', in E. Sinn (ed.), *Last Half Century of Chinese Overseas: Comparative Perspectives* (Hong Kong, 1994).
7. *Nelson Evening Mail* (NZ), 17 January 1908, p.2.
8. *The Argus* (Melbourne), 4 January 1908, p.14.
9. *Feilding Star*, 8 February 1908, p.2; *Otago Daily Times*, 7 March 1908, p.9; *The Sydney Morning Herald* (NSW), 2 September 1907, p.7.
10. *Nelson Evening Mail*, 17 January 1908, p.2.
11. Reckner, J. R., *Teddy Roosevelt's Great White Fleet: The World Cruise of the American Battlefleet, 1907–1909* (1988); Reckner, 'The Great White Fleet in

New Zealand', *Naval History*, 5:3 (Autumn, 1991); Wimmel, K., *Theodore Roosevelt and the Great White Fleet: American Sea Power Comes of Age* (1998).
12. *Manawatu Standard*, 16 January 1908, report of a meeting of the local Anti-Asiatic League, p.7; *Wairarapa Daily Times*, local meeting of the Anti-Chinese League, 25 June 1907, p.6.
13. See, for instance, Hyslop, *Bain*; W. H. Andrews, *Class Struggles in South Africa* (Cape Town, 1941); R. K. Cope, *Comrade Bill* (Cape Town, 1943).
14. London *Times*, 20 January 1904.
15. Samuel, pp.457–467; E. T. Cook, 'Liberal Colonial Policy', *Contemporary Review*, 91 (January/June 1907), pp.457–459.
16. See Rose, p.12.
17. R. Jebb, *Studies in Colonial Nationalism* (London, 1905), pp.131–132.
18. CO886/1/1, Dominions No.1, 'The Self-Governing Dominions and Coloured Immigration', memorandum by Sir Charles Lucas, head of Dominions Department, CO, July 1908; CO886/1/2, 'Suggestions as to Coloured Immigration into the Self-Governing Dominions', by Lucas, July 1908.
19. J. Ramsay MacDonald, *Labour and the Empire* (London, 1907), p.62; see also Neame, *The Asiatic Danger*, p.1.
20. Lake & Reynolds, p.233.
21. Thompson, 'Language', pp.151–152. For Curzon's views, see David Gilmour, *Curzon* (1994). For Milner's, see Lord Alfred Milner, *The Nation and the Empire* (London, 1913).
22. Offer, p.189.
23. See Cd.5745, Minutes of Proceedings of the Imperial Conference, 1911.
24. W. Wybergh, 'The Transvaal and the New Government', *Contemporary Review*, 89 (January/June 1906), p.315.
25. Richardson, *Chinese*, pp.42–43.

Bibliography

Secondary Sources

Accone, Darryl (2004), *All Under Heaven, The Story of the Chinese Family in South Africa* (Cape Town: David Philip).

Adhikari, Mohamed (September 2005), 'Contending Approaches to Coloured Identity and the History of the Coloured People of South Africa', *History Compass*, pp.1–16.

Akurang-Parry, Kwabena O. (2001), ' "We Cast about for a Remedy": Chinese Labor and African Opposition in the Gold Coast, 1874–1914', *The International Journal of African Historical Studies*, 34:2, pp.365–384.

Amery, L. S. (1953), *My Political Life, volume I* (London: Hutchinson).

Amery, Julian (1951), *Life of Joseph Chamberlain, volume IV* (London: Macmillan).

Anderson, Warwick (2006), *The Cultivation of Whiteness: Science, Health and Racial Destiny in Australia* (Durham, NC: Melbourne University Press).

Auerbach, Sascha (2009), *Race, Law, and "The Chinese Puzzle" in Imperial Britain* (Basingstoke, Hampshire: Palgrave Macmillan).

Bailey, Paul J. (2011), ' "An Army of Workers" Chinese Indentured Labour in First World War France', in S. Das (ed.), *Race, Empire and First World War Writing* (Cambridge: Cambridge University Press), pp.35–52.

Bailey, Paul J. (2009), 'Chinese Contract Workers in World War One: The Larger Context', in Zhang Jianguo (ed.), *Chinese Labourers and the First World War* (Jinan: Shandong University Press), pp.3–18.

Ballantyne, Tony (2002), *Orientalism, Race, Aryanism in the British Empire* (Basingstoke, Hampshire: Palgrave Macmillan).

Beaumont, Jacqueline (2004), 'An Irish Perspective on Empire: William Flavelle Monypenny', in Simon J Potter (ed.), *Newspapers and Empire in Ireland and Britain* (Portland, OR: Four Courts Press), pp.177–194.

Becker, Jules (1991), *The Course of Exclusion, 1882–1924, San Francisco Newspaper Coverage of the Chinese and Japanese in the United States* (San Francisco: Mellen Research University Press).

Beinart, William (September 1992), 'Introduction: Political and Collective Violence in Southern African Historiography', *Journal of Southern African Studies*, 18:3, pp.455–486.

Beloff, Max (1969), *Imperial Sunset, volume I, Britain's Liberal Empire, 1897–1921* (London: Methuen).

Benton, Gregor & Edmund Terence Gomez (2008), *The Chinese in Britain, 1800–Present, Economy, Transnationalism, Identity* (Basingstoke, Hampshire: Palgrave Macmillan).

Bickford-Smith, Vivian (1995), *Ethnic Pride and Racial Prejudice in Victorian Cape Town, Group Identities and Social Practice, 1875–1902* (Cambridge: Cambridge University Press).

Bickford-Smith, Vivian (1999), E. Van Heyningen & Nigel Worden, *Cape Town in the Twentieth Century: An Illustrated Social History* (Cape Town: David Philip).
Biko, Steve (1978), *I Write What I Like* (Chicago: University of Chicago Press).
Blue, Gregory (1999), 'Gobineau on China: Race Theory, the "Yellow Peril,"' and the Critique of Modernity', *Journal of World History*, 10:1, pp.93–134.
Bolt, Christine (1971), *Victorian Attitudes to Race* (London: Routledge).
Bosco, A. & A. May (eds.) (1997), *The Round Table: The Empire/Commonwealth and British Foreign Policy* (London: Lothian Foundation Press).
Boyce, D. George (ed.) (1990), *The Crisis of British Power: The Imperial and Naval Papers of the Second Earl of Selborne, 1895–1910* (London: Historians' Press).
Boyce, Robert W. D. (February 2000), 'Imperial Dreams and National Realities: Britain, Canada and the Struggle for a Pacific Telegraph Cable, 1879–1902', *The English Historical Review*, 115:460, pp.39–70.
Bozzoli, Belinda (1981), *The Political Nature of a Ruling Class: Capital and Ideology in South Africa, 1890–1933* (London: Routledge).
Bozzoli, Belinda (ed.) (1983), *Town and Countryside in the Transvaal: Capitalist Penetration and Popular Response* (Johannesburg: Ravan Press).
Breckenridge, K. (1998), 'The Allure of Violence: Men, Race and Masculinity on the South African Goldmines, 1900–1950', *Journal of Southern African Studies*, 24:4, pp.669–693.
Bridge, Carl & Kent Fedorowich (2003), 'Mapping the British World', *Journal of Imperial and Commonwealth History*, 31:2, pp.1–15.
Bridge, Carl & Bernard Attard (eds.) (2000), *Between Empire and Nation: Australia's External Relations from Federation to the Second World War* (Melbourne: Australian Scholarly Publishing).
Bright, Rachel (2013), 'Asian Migration and the British World, 1850–1914', in Andrew S. Thompson & Kent Fedorowich (eds.), *Empire, Identity and Migration in the British World* (Manchester: Manchester University Press).
Brookes, Edgar H. & Colin de B. Webb (1987), *A History of Natal* (Pietermaritzburg, 1965, 2nd edn: University of Natal Press).
Butler, Jeffrey & Deryck Schreuder (1987), 'Liberal Historiography Since 1945', in Jeffrey Butler, Richard Elphick, & David Welsh (eds.), *Democratic Liberalism in South Africa: Its History and Prospect* (Middletown, CN; Wesleyan University Press), pp.148–165.
Cain, P. & Hopkins A. (2001), *British Imperialism: 1688–2000* (London: Longman).
Cain, P. (2002), 'British Radicalism, the South African Crisis, and the Origins of the Theory of Financial Imperialism', in Omissi & Thompson (eds.), *The Impact of the South African War* (Basingstoke, Hampshire: Palgrave Macmillan), pp.173–193.
Cammack, Diana (1990), *The Rand at War, 1899–1902, The Witwatersrand & the Anglo-Boer War* (London: University of California Press).
Campbell, P. C. (1923), *Chinese Coolie Emigration to Countries within the British Empire* (London: King).
Cell, John W. (1982), *The Highest Stage of White Supremacy. The Origins of Segregation in South Africa and the American South* (London: Cambridge University Press).
Chanock, Martin (2001), *The Making of South African Legal Culture, 1902–1936: Fear, Favour, and Prejudice* (Cambridge: Cambridge University Press).

Chen, Ta (1967), *Chinese Migrations, with Special Reference to Labor Conditions* (Taipei, 1923).
Ching-Hwang, Yen (1985), *Coolies and Mandarins* (Singapore: Singapore University Press).
Churchill, Randolph S. (1967), *Winston S. Churchill, volume II, Young Statesman* (London: Heinemann).
Cohen, Robin (1980), 'Resistance and Hidden Forms of Consciousness amongst African Workers', *Review of African Political Economy*, 19, pp.8–22.
Cohen, S. (1972), *Folk Devils and Moral Panics: The Creation of the Mods and Rockers* (London: MacGibbon & Kee).
Cole, Douglas (May 1971), 'The Problem of "Nationalism" and "Imperialism" in British Settlement Colonies', *The Journal of British Studies*, 10:2, pp.160–182.
Colvin, Ian (1923), *The Life of Jameson, volume II* (London: E. Arnold and Company).
Cooper, Frederick (1980), *From Slaves to Squatters: Plantation Labor and Agriculture in Zanzibar and Coastal Kenya, 1890–1925* (New Haven: Yale University Press).
Cooper, Frederick (December 1994), 'Conflict and Connection: Rethinking Colonial African History', *The American Historical Review*, 99:5, pp.1516–1545.
Cooper, Frederick, Holt, Thomas Cleveland, & Scott, Rebecca Jarvis (2000), *Beyond Slavery: Explorations of Race, Labor, and Citizenship in Postemancipation Societies* (Chapel Hill, NC: University of North Carolina Press).
Cooper, Frederick (2005), *Colonialism in Question: Theory, Knowledge, History* (Berkeley: University of California Press).
Cornwell, Gareth (1996), 'George Webb Hardy's "The Black Peril" and the Social Meaning of Black Peril in Early Twentieth Century South Africa', *Journal of Southern African Studies*, 22, pp.441–453.
Crush, Jonathan, Alan Jeeves, & David Yudelman (1991), *South Africa's Labor Empire: A History of Black Migrancy to the Gold Mines* (Oxford: Oxford University Press).
Cuthbertson, G., A. Grundlingh, & M. Suttie (eds.) (2001), *Writing a Wider War: Rethinking Gender, Race, and Identity in the South African War, 1899–1902* (Cleveland, OH: Ohio University Press).
Darwin, John (2001), 'A Third British Empire? The Dominion Idea in Imperial Politics', in Judith M. Brown and Wm. Roger Louis (eds.), *The Oxford History of the British Empire, volume IV: The Twentieth Century* (Oxford: Oxford University Press), pp.64–87.
Davey, Arthur (1978), *The British Pro-Boers, 1877–1902* (Cape Town: Tafelberg).
Davenport, T. R. H. (1966), *The Afrikaner Bond: The History of a South African Political Party, 1880–1911* (London: Oxford University Press).
Davenport, T. R. H. (1987), *South Africa: A Modern History* (London, 1977, 3rd edn: Palgrave Macmillan).
Davies, Robert (1976–1977), 'Mining Capital, the State and Unskilled White Workers in South Africa 1901–1913', *Journal of Southern African Studies*, 3:1, pp.41–69.
Dawe, Richard D. (1998), *Cornish Pioneers in South Africa: 'Gold and Diamonds, Copper and Blood'* (Cornwall: Cornish Hillside Publications).
Denoon, D. J. N. (1968), 'Capitalist Influence and the Transvaal Government During the Crown Colony Period. 1900–1906', *The Historical Journal*, 11:2, pp.301–331.

Denoon, D. J. N. (March 1980), 'Capital and Capitalists in the Transvaal in the 1890s and 1900s', *Historical Journal*, 23, pp.111–132.
Denoon, D. J. N. (1973), *A Grand Illusion: The Failure of Imperial Policy in the Transvaal Colony during the Period of Reconstruction: 1900–1905* (London: Longman).
Doxey, G. V. (1961), *The Industrial Colour Bar in South Africa* (Cape Town: Greenwood Press).
Drew, Allison (2007), *Between Empire and Revolution: A Life of Sidney Bunting, 1873–1936* (London: Pickering & Chatto).
Dubow, Saul (2006), *A Commonwealth of Knowledge: Science, Sensibility and White South Africa, 1820–2000* (Oxford: Oxford University Press).
Dubow, Saul (1997), 'Colonial Nationalism, the Milner Kindergarten and the Rise of "South Africanism"': 1902–1910', *History Workshop Journal*, 43, pp.53–86.
Dubow, Saul (1995), *Scientific Racism in Modern South Africa* (Cambridge: Cambridge University Press).
Duminy, A. H. & W. R. Guest (eds.) (1976), *Fitzpatrick, South African Politician, Selected Papers, 1888–1906* (Johannesburg: McGraw-Hill Book Co.).
Eddy, J. & D. Schreuder (eds.) (1988), *The Rise of Colonial Nationalism: Australia, New Zealand, Canada and South Africa First Assert Their Nationalities, 1880–1914* (Sydney: Allen & Unwin).
Ellis, Stephen (September 2003), 'Violence and History: A Response to Thandika Mkandawire', *The Journal of Modern African Studies*, 41:3, pp.457–475.
Emden, Paul H (1935), *Randlords* (London: Hodder & Stoughton).
Evans, J., P. Grimshaw, D. Philips, & S. Swain (2003), *Equal Subjects, Unequal Rights: Indigenous Peoples in British Settler Colonies, 12830–1910* (Manchester, Manchester University Press).
Etherington, Norman (1988), 'Natal's Black Rape Scares of the 1870s', *Journal of Southern African Studies*, 15, pp.36–53.
Fairbank, J., Katherine Frost Bruner, & Elizabeth MacLeod Matheson (eds.) (1975), *The I. G. In Peking, Letters of Robert Hart, volume II* (London: Harvard University Press).
Fanon, Frantz (1963), *The Wretched of the Earth* (New York: Grove Press).
Feldman, David (2007), 'Jews and the British Empire c.1900', *History Workshop Journal*, 63, pp.70–89.
Forman, Ross G. (2002), 'Randy on the Rand: Portuguese African Labor and the Discourse on "Unnatural Vice" in the Transvaal in the Early Twentieth Century', *Journal of the History of Sexuality*, 11:4, pp.570–609.
Fort, G. Seymour (1908), *Dr. Jameson* (London: Hurst and Blackett).
Foucault, M. (1994), 'Governmentality' (1978) in P. Rabinow & N. Rose (eds.), *The Essential Foucault: Selections from The Essential Works of Foucault, 1954–1984* (New York: New Press).
Fraser, Maryna & Alan Jeeves (ed.) (1977), *All that Glittered, Selected Correspondence of Lionel Phillips, 1890–1924* (Cape Town: Oxford University Press).
Fraser, Maryna (ed.) (1985), 'Edmund Bright Letters', *Johannesburg Pioneer Journals, 1888–1909* (Cape Town: Van Riebeeck Society).
Fredrickson, George (1981), *White Supremacy, A Comparative Study in American and Southern African History* (Oxford: Oxford University Press).
Freed, Louis F. (1949), *The Problem of European Prostitution in Johannesburg* (Johannesburg: Juta).

Bibliography

Furedi, Frank (2002), *Culture of Fear: Risk Taking and the Morality of Low Expectation* (London, Cassell).
Gandhi, M. K. (1960+), *Collected Works, volumes III and IV* (Delhi: Indian Ministry of Information and Broadcasting).
Gandhi, M. K. (1927, 1982), *An Autobiography or the Story of My Experiments with Truth* (Hamondsworth, Middlesex: Courier Dover).
Garrard, John A. (1971), *The English and Immigration, 1880–1910* (London).
Garson, N. G. (1966), 'Het Volk: The Botha-Smuts Party in the Transvaal, 1904–11', *Historical Journal*, 9:1, pp.101–132.
Geiger, Andrea (Winter/Spring 2007), 'Negotiating the Boundaries of Race and Class: Meiji Diplomatic Responses to North American Categories of Exclusion', *British Columbia Studies*, 156/157.
Gilman, Sander L. (1985), *Difference and Pathology, Stereotypes of Sexuality Stereotypes of Sexuality, Race and Madness* (Ithaca, NY: Cornell University Press).
Giliomee, Hermann (1987), 'The Beginning of Afrikaner Nationalism, 1870–1915, *South African Historical Journal*, 19, pp.115–142.
Giliomee, Hermann (2003), *The Afrikaners, Biography of a People* (London: Hurst C. & Co.).
Gorman, Daniel (2002), 'Wider and Wider Still?: Racial Politics, Intra-Imperial Immigration and the Absence of an Imperial Citizenship in the British Empire', *Journal of Colonialism and Colonial History*, 3:3, http://muse.jhu.edu/journals/journal_of_colonialism_and_colonial_history/v003/3.3gorman.html. Last accessed on 5 April 2010.
Gorman, Daniel (March 2004), 'Lionel Curtis, Imperial Citizenship, and the Quest for Unity', *The Historian*, 66:1, pp.67–96.
Gouter, David (2007), *Guarding the Gates: The Canadian Labour Movement and Immigration, 1872–1934* (Vancouver: University of British Columbia Press).
Grant, Kevin (2005), *A Civilised Savagery: Britain and the New Slaveries in Africa, 1884–1926* (London: Routledge).
Grundlingh, Albert (1991), ' "Protectors and Friends of the People"? The South African Constabulary in the Transvaal and the Orange River Colony, 1900–1908', in D. M. Anderson & D. Killingray (eds.), *Policing the Empire: Government, Authority, and Social Control, 1830–1940* (Manchester: Manchester University Press), pp.168–182.
Grundlingh, G. (2006), *The Dynamics of Treason, Boer Collaboration in the South African War of 1899–1902* (Pretoria: Protea Book House), English translation of *Die 'hendsoppers' en 'joiners'* (1979).
Gungwu, Wang (1991), *China and the Chinese Overseas* (Singapore: Eastern Universities Press).
Guterl, Matthew & Christine Skwiot (Winter 2005), 'Atlantic and Pacific Crossings: Race, Empire, and "the Labor Problem" in the Late Nineteenth Century', *Radical History Review*, 91, pp.40–61.
Halpern, Rick (March 2004), 'Solving the "Labour Problem": Race, Work and the State in the Sugar Industries of Louisiana and Natal, 1870–1910', *Journal of Southern African Studies*, 30:1, pp.19–40.
Hancock, W. K. (1962), *Smuts: The Sanguine Years 1870–1919* (Cambridge).
Hancock, W. K. & J. van der Poel (eds.) (1966), *Selections from the Smuts Papers, volume II* (Cambridge: Cambridge University Press).

Harries, P. (1994), *Work, Culture and Identity: Migrant Labourers in Mozambique and South Africa, 1860–1910* (Johannesburg: Pearson Education).
Harries, Patrick (September 1990), 'Symbols and Sexuality: Culture and Identity on the Early Witwatersrand Gold Mines', *Gender & History*, 2:3, pp.318–336.
Harris, Cheryl (1993), 'Whiteness as Property', *Harvard Law Review*, 106, pp.1707–1791.
Harris, K. L. (1996), 'Gandhi, the Chinese and Passive Resistance', in J. M. Brown & M. Prozesky (eds.), Gandhi and South Africa: Principles and Politics (New York, St. Martin's Press), pp.69–94.
Harris, Karen (2004), 'Private and Confidential: The Chinese Mine Labourers and "Unnatural Crime"', *South African Historical Journal*, 50, pp.115–133.
Harris, Karen (1998), 'The Chinese "South Africans": An Interstitial Community', in Wang Ling-Chi & Wang Gungwu (eds.), *The Chinese Diaspora, Selected Essays*, volume II (Singapore: Eastern Universities Press), pp.275–294.
Harris, Karen (2010), 'Sugar and Gold: Indentured Indian and Chinese Labour in South Africa', *Journal of Social Science*, 25, pp.147–158.
Harrison, Philip (2002), 'Reconstruction and Planning in the aftermath of the Anglo-Boer South African War: The Experience of the Colony of Natal, 1900–1910', *Planning Perspectives*, 17, pp.1–20.
Headlam, Cecil (ed.) (1931), *The Milner Papers, volume II, South Africa 1899–1905* (London: Cassell).
Henshaw, Peter & Ronald Hyam (2003), *The Lion and the Springbok: Britain and South Africa since the Boar War* (Cambridge: Cambridge University Press).
Hewison, Hope Hay (1989), *Hedge of Wild Almonds, South Africa, The Pro-Boers & the Quaker Conscience, 1890–1910* (London: Boydell & Brewer).
Hirson, Baruch (ed.) (1983), *South Africa: The War of 1899–1902 and the Chinese Labour Question* (Wakefield, microform).
Hobson, J. A. (1901), *The Psychology of Jingoism* (London: Grant Richards).
Hobson, J. A. (1902), *Imperialism: A Study* (London: J. Pott).
Hocking, Anthony (1986), *Randfontein Estates, The First 100 Years* (Bethulie, Orange Free State: Hollards).
Hoerder, Dirk (2002), *Cultures in Contact: World Migrations in the Second Millennium* (Durham, NC: Duke University Press).
Holt, Thomas C. (1992), *The Problem of Freedom: Race, Labor and Politics in Jamaica and Britain, 1832–1938* (Baltimore, MD: JHU Press).
Howe, Anthony (1997), *Free Trade and Liberal England, 1846–1946* (Oxford: Oxford University Press).
Human, Linda (1984), *The Chinese People of South Africa: Freewheeling on the Fringes* (Pretoria: University of South Africa).
Humphriss, Deryk & David G. Thomas (1986), *Benoni, Son of my Sorrow: The Social, Political and Economic History of a South African Gold Mining Town* (Benoni: Benoni Town Council).
Huttenback, Robert A. (1976), *Racism and Empire: White Settlers and Coloured Immigrants in the British Self-Governing Colonies, 1830–1910* (Ithaca, NY: Cornell University Press).
Huttenback, Robert A. (November 1973), 'The British Empire as a "White Man's Country" – Racial Attitudes and Immigration Legislation in the Colonies of White Settlement', *Journal of British Studies*, 13:1, pp.108–137.

Huttenback, Robert A. (1971), *Gandhi in South Africa, British Imperialism and the Indian Question, 1860–1914* (London: Cornell University Press).
Huynh, Tu T. (May 2008), 'From Demand for Asiatic Labor to Importation of Indentured Chinese Labor: Race Identity in the Recruitment of Unskilled Labor for South Africa's Gold Mining Industry, 1903–1910', *Journal of Chinese Overseas*, 4:1, pp.51–68.
Hyam, Ronald (1968), *Elgin and Churchill at the Colonial Office, 1905–1908: The Watershed of the Empire-Commonwealth* (New York: Macmillan).
Hyam, Ronald (1965), 'Smuts and the Decision of the Liberal Government to Grant Responsible Government to the Transvaal, January and February 1906', *The Historical Journal*, 8:3, pp.380–398.
Hyam, Ronald (1991), *Empire and Sexuality: The British Experience* (Manchester: Manchester University Press).
Hyslop, Jonathan (1999), 'The Imperial Working Class Makes Itself "White": White Labourism in Britain, Australia, and South Africa Before the First World War', *Journal of Historical Sociology*, 12:4, pp.398–421.
Hyslop, Jonathan (2004), *The Notorious Sydicalist: J. T. Bain: A Scottish Rebel in Colonial South Africa* (Johannesburg: Jacana).
Irick, Robert (1982), *Ch'ing Policy Toward the Coolie Trade 1847–1878* (Taipei: Chinese Materials Center).
Jeeves, Alan (1985), *Migrant Labour in South Africa's Mining Economy* (Witwatersrand: McGill-Queen's Press).
Jensen, Gary F. (2007), *The Path of the Devil: Early Modern Witch Hunts* (Lanham, MD: Rowman & Littlefield Publishers).
Johnstone, Frederick A. (1976), *Class, Race and Gold: A Study of Class Relations and Racial Discrimination in South Africa* (London: University Press of America).
Johnston, Hugh J. M. (1979), *The Voyage of the Komagata Maru: The Sikh Challenge to Canada's Colour Bar* (Delhi: UBC Press).
Jung, Moon-Ho (2005), 'Outlawing "Coolies": Race, Nation, and Empire in the Age of Emancipation', *American Quarterly*, 57:3, pp. 677–701.
Kale, Madhavi (2010), 'Race and Empire: The Case of Indian Indentured Migration', at http://www.empire.amdigital.co.uk.libproxy.kcl.ac.uk/essays/content/MadhaviKale2.htm. Last accessed on 5 April 2010.
Katz, Elaine N. (June 1974), 'White Workers' Grievances and the Industrial Colour Bar, 1902–13', *South African Journal of Economics*, 42:2, pp.84–105.
Katz, Elaine N. (1976), *A Trade Union Aristocracy: A History of White Workers in the Transvaal and the General Strike of 1913* (Johannesburg: Wits).
Katz, Elaine N. (1995), 'The Underground Route to Mining – Afrikaners and the Witwatersrand Gold Mining Industry from 1902 to the 1907 Miner's Strike', *Journal of African History*, 36:3, pp.467–489.
Katzenellenbogen, Simon (1982), *South African and Southern Mozambique: Labour, Railways and Trade in the Making of the Relationship* (Manchester: Manchester University Press).
Keegan, Timothy J. (1987), *Rural Transformations in Industrialising South Africa: The Southern Highveld to 1914* (London: Macmillan).
Keegan, Timothy (1996), *Colonial South Africa and the Origins of the Racial Order* (Cape Town: Continuum).
Keith, Arthur B. (1912), *Responsible Government in the Dominions, volume II* (Oxford: Oxford University Press).

Kendle, J. E. (1967), *The Colonial and Imperial Conferences 1887–1911. A Study in Imperial Organisation* (London: Royal Commonwealth Society).

Kennedy, Brian (1984), *A Tale of Two Mining Cities. Johannesburg and Broken Hill: 1885–1925* (Johannesburg: Ad. Donker).

Kennedy, Dane (1987), *Islands of White: Settler Society and Culture in Kenya and Southern Rhodesia, 1890–1939* (Durham, NC: University of North Carolina Press).

Kesner, Richard M. (Spring 1978), 'The Transvaal, the Orange River Colony, and the South African Loan and War Contribution Act of 1903', *Albion: A Quarterly Journal Concerned with British Studies*, 10:1, pp.28–53.

Kiernan, V. G. (1969), *The Lords of Human Kind: European Attitudes Towards the Outside World in the Imperial Age* (London: Columbia University Press).

King, Michelle T. (2009), 'Replicating the Colonial Expert: The Problem of Translation in the Late Nineteenth-century Straits Settlements', *Social History*, 34:4, pp.428–446

Kirk, Joyce F. (June 1991), 'A "Native" Free State at Korsten: Challenge to Segregation in Port Elizabeth, South Africa, 1901–1905', *Journal of Southern African Studies*, 17:2, pp.309–336.

Krikler, Jeremy (2005), *White Rising: The 1922 Insurrection and Racial Killing in South Africa* (Manchester: Manchester University Press).

Krikler, Jeremy (1993), 'Social Neurosis and Hysterical Pre-cognition in South Africa: A Case Study and Reflections'. *South African Historical Journal*, 28, pp.491–520.

Kubicek, R. V. (1979), *Economic Imperialism in Theory and Practice: The Case of South African Gold Mining Finance, 1886–1914* (Durham, NC: Duke University Center for Commonwealth and Comparative Studies).

Kubicek, R. V. (1969), *The Administration of Imperialism: Joseph Chamberlain at the Colonial Office* (Durham, NC: Duke University Center for Commonwealth and Comparative Studies).

Kubicek, R. V. (1999), 'Economic Power at the Periphery: Canada, Australia and South Africa, 1850–1914', in Raymond E. Dumett (ed.), *Gentlemanly Capitalism and British Imperialism, The New Debate on Empire* (New York: Longman), pp.113–126.

Kynaston, David (1995), *The City of London, vol.II, Golden Years, 1890–1914* (London: Random House).

Kynoch, Gary (2003), 'Controlling the Coolies: Chinese Mineworkers and the Struggle for Labor in South Africa, 1904–1910', *The International Journal of African Historical Studies*, 36:2, pp.309–329.

Kynoch, Gary (September 2005), ' "Your Petitioners are in Mortal Terror": The Violent World of Chinese Mineworkers in South Africa, 1904–1910', *Journal of Southern African Studies*, 31:3, pp.531–546.

Lahiri, Shompa (2000), *Indians in Britain: Anglo-Indian Encounters, Race and Identity, 1880–1930* (London: Taylor & Francis).

Lai, Walton Look (1993), *Indentured Labor, Caribbean Sugar: Chinese and Indian Migrants to the British West Indies, 1838–1918* (Baltimore, MD: Johns Hopkins University Press).

Lake, Marilyn (2004), 'The White Man under Siege: New Histories of Race in the Nineteenth Century and the Advent of White Australia', *History Workshop Journal*, 58, pp.41–62.

Bibliography

Lake, Marilyn (2004), 'Translating Needs into Rights: The Discursive Imperative of the Australian White Man, 1901–30', in Stefan Dudink, Karen Hagemann, & John Tosh (eds.), *Masculinities in Politics and War: Gendering Modern History* (Manchester: Manchester University Press), pp.199–219.

Lake, Marilyn & Henry Reynolds (2008), *Drawing the Global Colour Line: White Men's Countries and the International Challenge of Racial Equality* (Cambridge: Cambridge University Press).

Lal, Brij V. (1985), 'Kunti's Cry: Indentured Women on Fiji Plantations', *Indian Economic Social History Review*, 22:55, pp.55–71.

Lambert, John (2006), ' "The Thinking is done in London": South Africa's English Language Press and Imperialism', in C. Kaul (ed.), *Media and the British Empire* (Basingstoke, Hampshire: Palgrave Macmillan), pp.37–54.

Legassick, M. (1995), 'British Hegemony and the Origins of Segregation in South Africa, 1901–14', in W. Beinart & S. Dubow (eds.), *Segregation and Apartheid in Twentieth Century South Africa* (London: Routledge) pp.43–59.

Lester, Alan (2001), *Imperial Networks: Creating Identities in Nineteenth-century South Africa and Britain* (London: Routledge).

Levy, N. (1982), *The Foundations of the South African Cheap Labour System* (London: Routledge & Kegan Paul).

Lewsen, P. (ed.) (1969), *Selections from the Correspondence of John X. Merriman, volume III, 1905–1924* (Cape Town: Van Riebeeck Society).

Lewsen, P. (1982), *John X. Merriman, Paradoxical South African Statesman* (New Haven, CT: Yale University Press).

Lockhart, J. G. & C. M. Woodhouse (1963), *Rhodes* (New York: Macmillan).

Long, B. K. (1941), *Drummond Chaplin* (London: Oxford university press).

Lowry, Donal (ed.) (2000), *The South African War Reappraised* (Manchester: Manchester University Press).

Macdonald, Andrew (May 2008), 'In the Pink of Health or the Yellow of Condition?: Chinese Workers, Colonial Medicine and the Journey to South Africa, 1904–1907', *Journal of Chinese Overseas*, 4:1, pp.23–50.

MacKenzie (1984), John, *Propaganda and Empire: The Manipulation of British Public Opinion 1880–1960* (Manchester: MUP).

Mackerras, Colin (2000), *Sinophiles and Sinophobes, Western Views of China* (Oxford: OUP).

Magubane, B. M. (1996), *The Making of a Racist State: British Imperialism and the Union of South Africa, 1875–1910* (Trenton, NJ: Africa World Press).

Magubane, Zine (2004), *Bringing the Empire Home: Race, Class, and Gender in Britain and Colonial South Africa* (Chicago: University of Chicago Press).

Malherbe, V. C. (1991), 'Indentured and Unfree Labour in South Africa: Towards an Understanding', *South African Historical Journal*, 24, pp.3–30.

Manning, Patrick (2005), *Migration in World History* (New York: Routledge).

Mansergh, N. (1982), *The Commonwealth Experience, Volume I, the Durham Report to the Anglo-Irish Treaty* (London, 1969, 2nd edn: University of Toronto Press).

Maquarries, J. W. (ed.) (1962), *The Reminiscences of Sit Walter Stanford, volume II, 1885–1929* (Cape Town: Angus & Robertson).

Marais, J. S. (1957), *The Cape Coloured People, 1652–1937* (Johannesburg: Witwatersrand University Press).

Marks, Shula & Richard Rathbone (eds.) (1982), *Industrialisation and Social Change in South Africa: African Class Formation, Culture and Consciousness, 1870–1930* (London).

Marks, Shula & S. Trapido (1992), 'Lord Milner and the South African State Reconsidered', in M. Twaddle (ed.), *Imperialism, the State and the Third World* (London: British Academic Press), pp.80–94.

Markus, Andrew (1979), *Fear and Hatred, Purifying Australia & California 1850–1901* (Sydney: Hale & Iremonger).

Marlowe, J. (1976), *Milner-Apostle of Empire: Life of Alfred George Rt. Hon Viscount Milner of St. James' and Cape Town* (London: Hamish Hamilton).

Martens, Jeremy C. (June 2002), 'Settler Homes, Manhood and "Houseboys": An Analysis of Natal's Rape Scare of 1886', *Journal of Southern African Studies*, 28:2, pp.379–400.

Martens, Jeremy (2006), 'A Transnational History of Immigration Restriction: Natal and New South Wales, 1896–97', *Journal of Imperial and Commonwealth History*, 34:3, pp.323–344.

Martens, Jeremy (2008), 'Richard Seddon and Popular Opposition in New Zealand to the Introduction of Chinese Labour into the Transvaal, 1903–1904', *New Zealand Journal of History*, 42:2, pp.176–195.

Martin, Ged (1975), 'The Idea of "Imperial Federation"', in Ronald Hyam & Ged Martin (eds.), *Reappraisals in British Imperial History* (London: Macmillan), pp. 121–138.

Mathur, Saloni (1999), 'Wanted Native Views: Collected Colonial Postcards of India', in A. Burton (ed.), *Gender, Sexuality, and Colonial Modernities* (London: Routledge), pp.95–116.

Maud, John P. R. (1938), *City Government: The Johannesburg Experiment* (Oxford).

Mawby, Arthur (2000), *Gold Mining and Politics – Johannesburg 1900–1907: The Origins of the Old South Africa? volume I & II* (Lewiston: Edwin Mellen Press).

Mawby, Arthur (1974), 'Capital, Government and Politics in the Transvaal, 1900–1907: A Revision and a Reversion', *Historical Journal*, 17:2, pp.387–415.

Le May, G. H. L. (1965), *British Supremacy in South Africa, 1899–1907* (Oxford: Oxford University Press).

McCulloch, Jock (2000), *Black Peril, White Virtue: Sexual Crime in Southern Rhodesia, 1902–1935* (Bloomington, IN: Indiana University Press).

McKeown, A. (May 1999), 'Conceptualizing Chinese Diasporas, 1842 to 1949', *The Journal of Asian Studies*, 58:2, pp.306–337.

McKeown, A. (2004), 'Global Migration 1846–1940', *Journal of World History*, 19:2, pp.155–189.

McKeown, A. (2008), *Melancholy Order: Asian Migration and the Globalization of Borders* (New York: Columbia University Press).

Merrington, Peter (1997), 'Masques, Monuments and Masons: The 1910 Pageant of the Union of South Africa', *Theatre Journal*, 49:1, pp.1–14.

Miller, Carman (1993), *Painting the Map Red: Canada and the South African War, 1899–1902* (Montreal: McGill-Queen's Press).

Miller, Stuart Creighton (1969), *The Unwelcome Immigrant: The American Image of the Chinese, 1785–1882* (Berkeley: University of California Press).

Mills, Catherine (2008), 'The Emergence of Statutory Hygiene Precautions in the British Mining Industries, 1890–1914', *The Historical Journal*, 51:1, pp.145–168.

Moody, T. Dunbar (1988), 'Migrancy and Male Sexuality on the South African Gold Mines', *Journal of Southern African Studies*, 14:2, pp.228–256.

Morgan, Kenneth O. (2002), 'The Boer War and the Media 1899–1902', *Twentieth Century British History*, 13:1, pp.1–16.

Murphy, Erin L. (2005), ' "Prelude to imperialism": Whiteness and Chinese Exclusion in the Reimagining of the United States', *Journal of Historical Sociology,* 18:4, pp.457–490.
Murray, B. K. (1982), *Wits: The Early Years* (Johannesburg: University of Witwatersrand Press).
Mutanbirwa, James A. Chamunorwa (1980), *The Rise of Settler Power – in Southern Rhodesia (Zimbabwe), 1898–1923* (London: Fairleigh Dickinson University Press).
Nasson, Bill (2004), 'Why They Fought: Black Cape Colonists and Imperial Wars, 1899–1918', *The International Journal of African Historical Studies,* 37:1, pp.55–70.
Ngcongco, L. D. (1979), 'John Tengo Jabavu, 1859–1921', in Christopher Saunders (ed.), *Black Leaders in South African History* (Guildford: Heinemann Educational), pp.142–56.
Nimocks, W. (1968), *Milner's Young Men: The "Kindergarten" In Edwardian Imperial Affairs* (London: Duke University Press).
Northrup, D. (1995), *Indentured Labour in the Age of Imperialism, 1834–1922* (Cambridge: Cambridge University Press).
Odendaal, A. (1984), *Vukani Bantu! The Beginnings of Black Protest Politics in South Africa to 1912* (Cape Town: Philip).
Offer, Avner (1989), *The First World War: An Agrarian Interpretation* (Oxford: Oxford University Press).
Omissi, D. E. & A. S. Thompson (eds.) (2002), *The Impact of the South African War* (Basingstoke, Hampshire: Palgrave Macmillan).
Otte, T. G. (2006), *The China Question: Great Power Rivalry and British Isolation, 1894–1905* (Oxford: Oxford University Press).
Ownby, David (June 2001), 'Recent Chinese Scholarship on the History of Chinese Secret Societies', *Late Imperial China,* 22:1, pp.139–158.
Packard, Randall M. (1993), 'The Invention of the "Tropical Worker": Medical Research and the Quest for Central African Labor on the South African Gold Mines, 1903–1936', *The Journal of African History,* 34:2, pp.271–292.
Page, R. J. D. (February 1970), 'Canada and the Imperial Idea in the Boer War Years ', *Journal of Canadian Studies,* 5:1, pp.33–49.
Pellew, Jill (June 1989), 'The Home Office and the Aliens Act, 1905', *The Historical Journal,* 32:2, pp.369–385.
Pelling, Henry (1968), *Popular Politics and Society in Late Victorian Britain* (London: Palgrave Macmillan).
Phimister, Ian (1988), *An Economic and Social History of Zimbabwe, 1890–1948: Capital Accumulation and Class Struggle* (London: Longman).
Phimister, Ian (2006), 'Foreign Devils, Finance and Informal Empire: Britain and China c.1900–1912', *Modern Asian Studies,* 40:3, pp.737–759.
Pillay, Bala (1976), *British Indians in the Transvaal. Trade, Politics, and Imperial Relations, 1865–1906* (London: Longman).
Pineo, Huguette Li-Tio-Fane (1985), *Chinese Diaspora in Western Indian Ocean* (Singapore: Ed. de l'océan indien), translation from French.
Pocock, J. G. A. (1975), 'British History: A Plea for a New Subject', *Journal of Modern History,* 47:4, pp.601–621.
Porter, A. N. (June 1973), 'Sir Alfred Milner and the Press, 1897–1899', *The Historical Journal,* 16:2, pp.323–339.

Porter, A. (1980), *The Origins of the South African War: Joseph Chamberlain and the Diplomacy of Imperialism, 1895–99* (Exeter: Manchester University Press).

Potter, Simon J. (2003), *News and the British World: The Emergence of an Imperial Press System, 1876–1922* (Oxford: Oxford University Press).

Potter, Simon J. (2007), 'Richard Jebb, John S. Ewart and the Round Table, 1898–1926', *English Historical Review*, 122:495, pp.105–132.

Potts, Lydia (1990), *The World Market, A History of Migration* (London: Zed Books), translation by Terry Bond.

Pyrah, G. B. (1955), *Imperial Policy and South Africa. 1902–1910* (Oxford: Oxford University Press).

Raynard, J. H. (1940), *Dr. A Abdurahman: A Biographical Memoir* (Cape Town: Friends of the National Library of South Africa).

Read, Donald (1996), 'Reuters and South Africa: "South Africa is a country of monopolies"', *South African Journal of Economic History*, 11:2, pp.104–143.

Readman, Paul (2001), 'The Liberal Party and Patriotism in Early Twentieth Century Britain', *Twentieth Century British History*, 12:3, pp.269–302.

Reckner, J. R. (1988), *Teddy Roosevelt's Great White Fleet: The World Cruise of the American Battlefleet, 1907–1909* (Annapolis, MD: Naval Institute Press).

Reckner, R. (Autumn, 1991), 'The Great White Fleet in New Zealand', *Naval History*, 5:3, pp.26–29.

Reinders, Eric (2004), *Borrowed Gods and Foreign Bodies: Christian Missionaries Imagine Chinese Religion* (London: University of California Press).

Richardson, Peter (1976), 'Coolies and Randlords: The North Randfontein Chinese Miners' "Strike" of 1905', *Journal of Southern African Studies*, 2:2, pp.151–177.

Richardson, Peter (1977), 'The Recruiting of Chinese Indentured Labour for the South African Gold-Mines, 1903–8', *Journal of African History*, 18:1, pp.85–107.

Richardson, Peter & Jean Jacques Van-Helten (1982), 'Labour in the South African Gold Mining Industry, 1886–1914', in Shula Marks & Richard Rathbone (eds.), *Industrialisation and Social Change in South Africa: African Class Formation, Culture and Consciousness, 1870–1930* (London: Longman), pp.77–98.

Richardson, Peter (1982), *Chinese Mine Labour in the Transvaal* (London: Macmillan).

Richardson, Peter (1984), 'Chinese Indentured Labour in the Transvaal Gold Mining Industry, 1904–1910', in Kay Saunders (ed.), *Indentured Labour in the British Empire, 1834–1920* (London: Croom Helm), pp.260–290.

Richardson, P. (1986), 'The Natal Sugar Industry in the Nineteenth Century', in W. Beinart, P. Delius, & S. Trapido (eds.), *Putting a Plough to the Ground: Accumulation and Dispossession in Rural South Africa* (Johannesburg: Ravan Press), pp.129–175.

Roediger, David R. (1991), *The Wages of Whiteness: Race and the Making of the American Working Class* (New York: Verso).

Rosenthal, E. (1970), *Gold! Gold! Gold! The Johannesburg Gold Rush* (New York: Macmillan).

Ross, Robert (1993), *Beyond the Pale: Essays on the History of Colonial South Africa* (Johannesburg: Wits University Press).

Russell, A. K. (1973), *Liberal Landslide. The General Election of 1906* (Newton Abbot: Elliots Books).

252 Bibliography

Sacks, Benjamin (1970), *South Africa: an Imperial Dilemma-Non-Europeans and the British Nation-1902–1914* (Albuquerque, NM: University of New Mexico Press).
Samson, Jane (ed.) (2005), *Race and Empire* (London: Longman).
Saxton, Alexander (1975), *The Indispensible Enemy, Labor and the Anti-Chinese Movement in California* (London: University of California Press).
Seed, John (2006), 'Limehouse Blues: Looking for Chinatown in the London Docks, 1900–1940', *History Workshop Journal*, 62:1, pp.58–85.
Shaw, Gerald (1999), *The Cape Times, An Informal History* (Cape Town: David Philip).
Schwarz, Bill (Autumn, 1998), 'Politics and Rhetoric in the Age of Mass Culture', *History Workshop Journal*, 46, pp.129–160.
Sinn, E. (ed.) (1994), *Last Half Century of Chinese Overseas: Comparative Perspectives* (Hong Kong: Hong Kong University Press).
Smith, I. (1995), *The Origins of the South African War, 1899–1902* (Harlow: Longman).
Solomon, Vivian (ed.) (1981), *Selections from the Correspondence of Percy Alport Molteno, 1892–1914* (Cape Town: Van Riebeeck Society).
Spencer, Scott C. (May 2010), 'British Liberty Stained: "Chinese Slavery," Imperial Rhetoric, and the 1906 British General Election', *Madison Historical Review*, 7, pp.1–27.
Stoler, A. & Cooper, F. (eds.) (1997), *Tensions of Empire: Colonial Cultures in a Bourgeois World* (Berkeley: University of California Press).
Summerskill, Michael Brynmôr (1982), *China on the Western Front: Britain's Chinese work force in the First World War* (London: Michael Summerskill).
Swan, M. (1985), *Gandhi, the South African Experience* (Johannesburg: Ravan Press).
Switzer, Les (1986), 'Gandhi in South Africa: The Ambiguities of Satyagraha', *Journal of Ethnic Studies*, 14:1, pp.122–128.
Tayal, Maureen (1977), 'Indian Indentured Labour in Natal, 1890–1911', *The Indian Economic and Social History Review*, 14:4, pp.519–547.
Temperley, Howard (ed.) (2000), *After Slavery: Emancipation and its Discontents* (London: Routledge).
Thompson, Andrew (2005), *The Empire Strikes Back? The Impact of Imperialism on Britain from the Mid-Nineteenth Century* (Harlow: Routledge).
Thompson, Andrew S. (April 1997), 'The Language of Imperialism and the Meaning of Empire: Imperial Discourse in British Politics, 1895–1914', *The Journal of British Studies*, 36:2, pp.147–177.
Thompson, Kenneth (1998), *Moral Panics* (London: Routledge).
Thompson, L. M. (1960), *The Unification of South Africa, 1902–1910* (Oxford: Oxford University Press).
Tinker, Hugh (1974), *A New System of Slavery: The Export of Indian Labour Overseas, 1830–1920* (London: Hansib).
Torrance, David E. (1996), *The Strange Death of the Liberal Empire: Lord Selborne in South Africa* (Liverpool: Liverpool University Press).
Trapido, Stanley (October 1978), 'Landlord and Tenant in a Colonial Economy: The Transvaal 1880–1910', *Journal of Southern African Studies*, 5, pp.26–58.
van Der Ross, R.E. (ed.) (1990), *Say it Out: The A.P.O Presidential Addresses and Other Major Political Speeches, 1906–1940 of Dr. A. Abdurahman* (Cape Town: University of the Western Cape).

Van-Helton, J. J. & Keith Williams (October 1983), ' "The Crying Need of South Africa": The Emigration of Single British Women to the Transvaal, 1901–10', *Journal of Southern African Studies*, 10, pp.17–38.
Van-Helten, J. J. & P. Richardson (1984), 'The Development of the South African Gold-Mining Industry 1895–1918', *The Economic History Review*, 37:3, pp.319–340.
van Heyningen, Elizabeth (1996), 'Leander Starr Jameson', in Jane Carruthers (ed.), *The Jameson Raid, A Centennial Retrospective* (Houghton, SA: Brenthurst Press), pp.181–192.
van der Horst, Sheila (1942), *Native Labour in South Africa* (London: Frank Cass).
van Onselen, Charles (1982), *Studies in the Social and Economic History of the Witwatersrand: 1886–1914, volume 1 New Babylon* and *volume II New Nineveh* (London: Longman).
van Onselen, Charles (February 1990), 'Race and Class in the South African Countryside: Cultural Osmosis and Social Relations in the Sharecropping Economy of the South Western Transvaal, 1900–1950', *The American Historical Review*, 95:1, pp.99–123.
Van Reenen, Rykie (ed.) (1984), *Emily Hobhouse: Boer War Letters* (Pretoria: Human & Rousseau).
Walker, David (2005), 'Australia's Asian Futures', in Martyn Lyons & Penny Russell (eds.), *Australia's History: Themes and Debates* (Sydney, UNSW Press), pp.63–80.
Walker, Ivan L. & Ben Weinbren (1961), *2000 Causalities, a History of the Trade Unions and the Labour Movement in the Union of South Africa* (Pietermaritzburg: South African Trade Union Council).
Warwick (1983), *Black People and the South African War, 1899–1902* (London: Cambridge University Press).
Warwick, Peter & S. B. Spies (eds.) (1980), *The South African War: The Anglo-Boer War 1899–1902* (Harlow: Longman).
Webber, H. O'Kelly (1936), *The Grip of Gold. The Life Story of a Dominion* (London: Hutchinson).
Webster, E. (ed.) (1978), *Essays in Southern African Labour History* (Johannesburg: Ravan).
Weinthal, Leo (ed.) (1929), *Memoirs, Mines & Millions, The Life Story of Sir Joseph Robinson BRT* (London: Simkin Marshall).
Wheatcroft, G. (1993), *The Randlords* (London: Weidenfeld).
Wilcox, Craig (2002), *Australia's Boer War: The War in South Africa 1899–1902* (London: Oxford University Press).
Wilson, Jon E. (2008), 'Agency, Narrative, and Resistance', in, Sara Stockwell (ed.), *The British Empire: Themes and Perspectives* (Oxford: Blackwell), pp.245–268.
Wimmel, K. (1998), *Theodore Roosevelt and the Great White Fleet: American Sea Power Comes of Age* (Washington: Brassey's Inc.).
Yap, Melanie & Dianne Leong Man (1996), *Colour, Confusion and Concessions: The History of the Chinese in South Africa* (Hong Kong: Hong Kong University Press).
Yarwood, A. T. (1967), *Asian Migration to Australia: The Background to Exclusion, 1896–1923* (Cambridge: Greenwood Publishing Group).
Yudelman, David (1975), 'Lord Rothschild, Afrikaner Scabs and the 1907 Strike: A State-Capital Daguerreotype', *African Affairs*, 74:294, pp.82–96.

Published primary sources

Anonymous (17 October 1902), *British Indians in the Transvaal and the Orange River Colonies*, reprinted from *India*.
Anonymous (February 1904), *Notes on the Labour Position in the Transvaal by the London Secretary of the Transvaal Chamber of Mines* (London).
Anti-Asiatic Importation League (January 1904), Letter, a pamphlet (Cape Town).
A.W. (1904), 'Yellow Slavery – and White', *Westminster Review*, 99, pp.477–491.
Bailey, Abe (1904), *Speeches on The Chinese Question by Mr. Abe Bailey, MLA' Delivered in the House of Assembly*, Cape Town, 8 and 16 March 1904. Reprinted from Cape Times (Cape Town).
Balfour Browne, John Hutton (1905), *South Africa, A Glance at Current Conditions and Politics* (London).
Barlow, George (1914), *The Poetical Works, 1902–1914* (London).
Beesly, E. S. (April 1904), 'Yellow Labour', *Positivist Review*, pp.139–155.
Birnbaum, Doris (June 1905), 'Chinese Labour in the Transvaal', *Independent Review*, pp.142–153.
Browne, R. E. (1907), 'Working Costs of the Mines of the Witwatersrand', *Journal of the South African Institution of Engineers*, 12, pp.332–333.
Bruce, M. C. (1908), *The New Transvaal* (London).
Buchan, John (1940), *Memory Hold the Door* (London).
Burns, John (May 1904), 'Slavery in South Africa', *Independent Review*, 2, pp.594–611.
Burns, John, MP (May 1904), 'Bondage for Black. Slavery for Yellow Labour.' Reprinted from *Independent Review* (London).
Burt, Thomas (1904), 'Chinese Labour: How Public Opinion is Ascertained, or Manufactured in the Transvaal.'
Burt, Thomas (April 1904), *Northumberland Miners' Mutual Confident Association Monthly Circular*.
Burt, Thomas (1905), *A Visit to the Transvaal* (Newcastle-upon-Tyne).
Chamber of Mines, *Annual Reports*, 1898, 1903, 1904, 1905, 1906, 1907, 1908, 1909, 1910.
Chaplin, F. D. P. (April 1903), 'The Labour Question in the Transvaal', *National Review*, pp.296–367.
Chaplin, F. D. P. (February 1905), 'The Labour Question in the Transvaal.' *National Review*, pp.999–1000.
Chilvers, Hedley A. (1929), *Out of the Crucible* (London).
Chilvers, Hedley A. (1933), *The Yellow Man Looks On* (London).
Churchill, W. S. (30 January 1906), *A Note upon the Transvaal Constitution Question*.
Clough, Ernest (1909), *The South African Parliamentary Manual* (London).
Conservative Publication Department (1904), *Chinese Labour, A Protest from Transvaal Nonconformist Ministers* (London).
Conservative Publication Department (1905), *Chinese Labour Brings More Employment for Whites in the Transvaal* (London).
The Consolidated Gold Fields of South Africa Ltd. (1937), *The History of the Gold Fields* (London).
Creswell, F. H. P. (1905), *The Chinese Labour Question from Within – Facts, Criticisms, & Suggestions – Impeachment of a Disastrous Policy* (London).

Creswell, F. H. P. (1906), *The Witwatersrand Gold Mines and Responsible Government: Cause of the Depression* (Cape Town).
Darragh, J. T. (January/June 1902), 'The Native Problem in South Africa', *Contemporary Review*, 81, p.97.
Dyer, Jerome E. (January/June 1903), 'The South African Labour Question', *Contemporary Review*, 83, pp.439–445.
Densmore, G. B. (1880), *The Chinese in California* (San Francisco).
Devitt, N (1934), *Memoirs of a Magistrate* (London).
Des Voeux, Sir W. (April 1906), 'Chinese Labour in the Transvaal: A Justification', *The Nineteenth Century and After*, pp.581–594.
Dickens the Younger, Charles (ed.) (1879), *Dickens's Dictionary of London: An Unconventional Handbook* (London).
Farrar, Sir George (1903), 'The South African Labour Problem. Speech by Sir George Farrar at a Meeting Held on the East Rand Proprietary Mines, On 31 March 1903' (London).
Fawcett, Millicent Garrett (July/December 1903), 'Impressions of South Africa, 1901 and 1903', *Contemporary Review*, 84.
Foster, Arnold (1905), 'Chinese Coolie Labour in South Africa: A Missionary's Rejoinder to the Rev. T. W. Pearce', *Chronicle of the London Missionary Society* (London), pp.19–23.
Fox, Alfred F., et al. (January/June 1903), 'The Native Labour Question in South Africa', *Contemporary Review*, 83, pp.542–543.
Fremantle, H. E. S. (July/December 1903), 'The Political Position at the Cape', *Contemporary Review*, 84, pp.526–539.
Goldman, Charles Sydney (May 1904), 'South Africa and Her Labour Problem', *Nineteenth Century & After*, 4.
Hales, Frank (July 1904), 'The Transvaal Labour Difficulties', *Fortnightly Review*, pp.110–123.
Harcourt, Rt Hon Sir William (1903), MP, *Native Labour in South Africa. Two Letters to the Times*, reprinted by the Liberal Publication Dept. (London).
Harris, John H. (1910), *Coolie Labour in the British Crown Colonies and Protectorates* (London).
Hobson, J. A. (January/June 1900), 'Capitalism and Imperialism in South Africa', *Contemporary Review*, 77, pp.1–17.
Imperial South Africa Association (ISAA) (1904), *Chinese Labour: Dignified Rebuke from Nonconformist Ministers in the Transvaal to Their Brethren in England* (London).
ISAA (1904), *Chinese Labour: Five Reasons for Supporting the Government on Chinese Labour* (London).
ISAA (1904), *Free Church Approval of Chinese Labour* (London).
ISAA (1904), *Chinese Labour for the Transvaal Mines: Nonconformists Condemn Agitation Against It* (London).
ISAA (1904), *Slavery in the Transvaal* (London).
ISAA (1904), *Voter Registration* (London).
ISAA (1904), *A Scottish South African MP on Transvaal Labour: Mr John Stroyan, MP, on the Importation of Chinese Labourers* (London).
ISAA (1904), *Chinese Labour. Five Reasons for Supporting the Government on Chinese Labour* (London).
ISAA, A. Jabez Strong (1904), *The Transvaal Labour Question. A British Workingman to British Workingmen. What It All Means to British Labourers*.

Innes, James Rose (1949), *Autobiography* (London).
Jebb, R. (1905), *Studies in Colonial Nationalism* (London).
Jones, Roderick (May 1905), 'The Black Problem in South Africa', *The Nineteenth Century and After*, pp.712–723.
Kinloch Cooke, C. (1904), *Chinese Labour (in the Transvaal), Being a Study of Its Moral, Economic, and Imperial Aspects*, reproduced from *The Empire Review* (London).
Kruger, D. W. (1969), *The Making of a Nation: A History of the Union of South Africa, 1910–1961* (London).
Leys, P. (January/June 1902), 'Chinese Labour for the Rand', *The Nineteenth Century and After*, 51, pp.183–194.
MacDonald, J. Ramsey (1902), *What I Saw in South Africa, September and October 1902* (London).
Macnamara, T. J. (1904), *Chinese Labour* (London).
Maddison, F. (1900), 'Why British Workmen Condemn the War', *North American Review*, 170:521, pp.518–528.
Markham, Violet (1913), *The South African Scene* (London).
Markham, Violet (1904), *The New Era in South Africa, with an Examination of the Chinese Labour Question* (London).
Mather, William (1904), *Chinese Workers on the Witwatersrand Mines* (Johannesburg).
Maxim, H. S. (March 1903), 'The Chinese and the South African Labour Question', *Fortnightly Review*, 73, pp.506–511.
Molteno, P. A. (1896), *A Federal South Africa* (London).
Morgan, Ben H. (20 November 1903–11 November 1904), 'Africa, (South), regeneration of', *Journal of the Society of Arts*, 52, pp.497–498.
Munro, A. E. (n.d.), *Transvaal Chinese Labour Difficulty*.
Naylor, T. (1904), *The Truth About the Chinese in South Africa* (London: London *Chronicle* Office).
Neame, L. E. (1907), *The Asiatic Danger in the Colonies* (London).
Pearce, Rev. T. W. (1904), 'Chinese Coolie Labour in South Africa, A Statement from the Chinese Missionary Standpoint', *Chronicle of the London Missionary Society*, pp.226–228.
Pease, Sir Alfred (July/December 1906), 'The Native Question in the Transvaal', *Contemporary Review*, 90, pp.16–36.
Perry, F. (1 November 1906), 'The Transvaal Labour Problem', *Paper Read to the Fortnightly Club* (Johannesburg).
Phillips, Lionel (1924), *Some Reminiscences* (London).
Phillips, Lionel (1905), *Transvaal Problems: Some Notes on Current Politics* (London).
Plaatje, Sol T. (1916), *Native Life in South Africa* (London).
Praagh, L. V. (ed.) (1906), *The Transvaal and its Mines* (London).
Rathbone, Edgar P. (September 1903), 'The Native Labour Question', *Nineteenth Century and After*, pp.404–413.
Rose, Edward B. (1906), *Uncle Tom's Cabin* (London).
Samuel, Herbert, MP (January/June 1904), 'The Chinese Labour Question', *Contemporary Review*, 85, pp.457–467.
Schumacher, R. W. (1906), *A Transvaal View of the Chinese Labour Question* (London).
Seely, J. E. B. (January 1904), 'Greater Britain: South Africa – White, Black and Yellow', *National Review*.

Seely, J. E. B. (1930), *Adventure* (London).
Stanley, P. Hyatt (November 1905–October 1906), 'The Black Peril in South Africa', *Macmillan's Magazine*, 1, pp.392–400.
Stead, W. T. (June 1904), 'South Africa and Its Problems: The Chinese Question', *Review of Reviews*, pp.37–46.
Thompson, H. C. (March 1906), 'Chinese Labour and Imperial Responsibility', *Contemporary Review*, pp.431–437.
Transvaaler (nom de plume for Geoffrey Robinson) (May 1905). 'Political Parties of the Transvaal', *National Review*, 45, pp.480–481.
Trew, H. F. (1937), *African Man Hunts* (London).
Whiteside, P. (19 June 1907), *Pamphlet of a Speech from the Debates of the Transvaal Legislative Assembly*.
Wybergh, W. (March 1906), 'The Transvaal and the New Government', *Contemporary Review*.
Wybergh, W. (May 1907), 'Imperial Organisation and the Colour Question. I', *Contemporary Review*, pp.695–705.
Wybergh, W. (June 1907), 'Imperial Organisation and the Colour Question. II', *Contemporary Review*, pp.805–815.
Wyndham, Hugh (1936), 'The Formation of the Union, 1901–1910', in E. A. Benians & A. P. Newton, *The Cambridge History of the British Empire, volume VIII* (London), pp.613–742.

Parliamentary papers

1844 [Cd.530] Correspondence relative to Emigration of Labourers to W. Indies and Mauritius, from W. Coast of Africa, E. Indies and China.
1897 [Cd.8596] Proceedings of Conference between Secretary of State for Colonies and Premiers of Self-governing Colonies, at Colonial Office, June and July 1897.
1903 [Cd.1463] Further Correspondences relating to Affairs in South Africa.
1903 [Cd1640] Minutes of Proceedings of the South African Customs Union Conference held at Bloemfontein, March 1903.
1903 [Cd.1552] Papers relating to the Finances of the Transvaal and Orange River Colony.
1903 [Cd.1586] Statement of the Estimated Financial position of the Transvaal and Orange River Colony.
1903 [Cd.1599] Draft Customs Union Convention agreed to by the Representatives of the British Colonies and Territories in South Africa at a Conference at Bloemfontein in March 1903.
1903 [Cd.1683] Correspondence relating to a proposal to employ Indian Coolies under Indenture on Railways in the Transvaal and Orange River Colonies.
1904 [Cd.1894] Reports of the Transvaal Labour Commission.
1904 [Cd.1895] Correspondence (containing the Chinese Labour Ordinance) relating to Affairs in the Transvaal and Orange River Colony.
1904 [Cd.1896] Reports of the Transvaal Labour Commission.
1904 [Cd.1897] Minutes of Proceedings and Evidence.
1904 [Cd.1898] Telegraphic Correspondence relating to the Transvaal Labour Ordinance, with Appendix the Ordinance amended in accordance with Telegrams.

258 Bibliography

1904 [Cd.1899] Telegraphic Correspondence relating to the Transvaal Labour Ordinance, with Appendix the Ordinance amended in accordance with Telegrams.
1904 [Cd.1950] Correspondence Relating to Recruiting of Labour in the British Central African Protectorate for Employment in the Transvaal, March 1904.
1904 [Cd.1956] Convention between the United Kingdom and China respecting the Employment of Chinese Labour in British Colonies and Protectorates, Signed at London, 13th May 1904. No.6.
1904 [Cd.2025] Correspondence relating to Conditions of Native Labour employed in Transvaal Mines.
1904 [Cd.2026] Telegraphic Correspondence relating to the Transvaal Labour Ordinance, with Appendix the Ordinance amended in accordance with Telegrams.
1904 [Cd.2028] Correspondence relating to the proposed introduction of Indentured Asiatic (Chinese) Labour into Southern Rhodesia.
1904 [Cd.2104] Correspondence (containing the Chinese Labour Ordinance) relating to Affairs in the Transvaal and Orange River Colony.
1904 [Cd.1941] Telegraphic Correspondence relating to the Transvaal Labour Ordinance, with Appendix the Ordinance amended in accordance with Telegrams.
1904 [Cd.1945] Telegraphic Correspondence relating to the Transvaal Labour Ordinance, with Appendix the Ordinance amended in accordance with Telegrams.
1904 [Cd.1956] Convention between the United Kingdom and China respecting the Employment of Chinese Labour in British Colonies and Protectorates, Signed at London, 13th May 1904.
1904 [Cd.1986] Telegraphic Correspondence relating to the Transvaal Labour Ordinance, with Appendix the Ordinance amended in accordance with Telegrams.
1904 [Cd.2183] Telegraphic Correspondence relating to the Transvaal Labour Ordinance, with Appendix the Ordinance amended in accordance with Telegrams.
1905 [Cd.2246] Convention between the United Kingdom and China respecting the Employment of Chinese Labour in British Colonies and Protectorates. Signed at London, 13 May 1904.
1905 [Cd.2401] Telegraphic Correspondence relating to the Transvaal Labour Ordinance, with Appendix the Ordinance amended in accordance with Telegrams.
1905 [Cd.2479] Papers relating to Constitutional Changes in the Transvaal.
1905 [Cd.2563] Telegraphic Correspondence relating to the Transvaal Labour Ordinance, with Appendix the Ordinance amended in accordance with Telegrams.
1906 [Cd.2786] Telegraphic Correspondence relating to the Transvaal Labour Ordinance, with Appendix the Ordinance amended in accordance with Telegrams.
1906 [Cd.2788] Telegraphic Correspondence relating to the Transvaal Labour Ordinance, with Appendix the Ordinance amended in accordance with Telegrams.

1906 [Cd.3025] Telegraphic Correspondence relating to the Transvaal Labour Ordinance, with Appendix the Ordinance amended in accordance with Telegrams.
1906 [Cd.3251] The Asiatic Law Amendment Ordinance No. 29 of 1906.
1907 [Cd.3308] Correspondence relating to Legislation affecting Asiatics in the Transvaal.
1907 [Cd.3338] Annual Report of the Foreign Labour Department of Johannesburg, 1905–1906.
1907 [Cd.3405] Correspondence relating to the Introduction of Chinese Labourers into the Transvaal in excess of the number of Licenses issued.
1907 [Cd. 3523] Minutes of Proceedings of the Colonial Conference, 1907.
1908 [Cd.3994] Correspondence relating to the Indentured Labour Laws Temporary Continuance Act, 1907.
1908 [Cd.4120] Statement of Mortality amongst the different classes of coloured labourers in the Transvaal Mines from January 1906 to February 1908.
1911 [Cd.5745] Imperial Conference (Minutes of Proceedings).

Newspapers & Journals

The Advertiser, Adelaide
Amalgamated Engineers' Journal
The Argus, Melbourne
Bloemfontein Post
The Brisbane Courier
Bulawayo Observer
The Cape Argus, Cape Town
The Cape Daily Telegraph, Port Elizabeth
The Cape Mercury, King William's Town
Cape Times, Cape Town
Daily News, London
De Volkstem, Pretoria
De Vriend/The Friend, Bloemfontein
East London Daily Dispatch and Frontier Advertiser
The Eastern Province Herald, Port Elizabeth
The East Rand Express, Germiston
Evening Post, Auckland
The Heidelberg News
Imvo Zabantsundu, King William's Town
Indian Opinion, Phoenix, Natal
Land en Volk, Pretoria
Manchester *Guardian*
Marlborough *Express*
The Mercury, Tasmania
Midland News, Cradock
The Natal Mercury, Durban
The Natal Witness, Pietermaritzburg
The New Zealand Herald
Otago Witness

The Owl, Cape Town
Poverty Bay Herald
Pretoria News
The Prince, Durban
Rand Daily Mail, Johannesburg
South African News, Cape Town
South African Typographical Journal
The Star, Johannesburg
The Sydney Morning Herald
The Times, London
Times of Natal
The Transvaal Advertiser, Pretoria
Transvaal Leader, Johannesburg
Tuapeka Times
Wanganui Chronicle

Unpublished Material

National archives, Kew
Public Record Office: Colonial Office (PRO CO)
Public Record Office: Foreign Office (PRO FO)

National archives repository, Pretoria
Archives of the Colonial Secretary (CS)
Archives of the Foreign Labour Department (FLD)
Archives of the Lieutenant-Governor (LtG)
Archives of the Law Department (LD)
Archives of the High Commissioner (HC)
Archives of the Prime Minister (PM)
Archives of the Secretary of Native Affairs (SNA)

Papers
Milner Papers, Bodleian Library, Oxford (MP)
Papers of William Waldegrave Palmer, 2nd Earl of Selborne, Bodleian Library, Oxford (SP)
Farrar Papers, Rhodes House, Oxford (FP)
Merriman Papers, Library of South Africa, Cape Town (MmM)
Archives of Herbert Eckstein & Co., Barlowe World Archives, Johannesburg (HE)

Theses and unpublished papers
Gordon, S. I. (1987), 'The Chinese Labour Controversy in British Politics and Policy-Making', University of Ulster Thesis.
Grey, P. C. (1969), 'The Development of the Gold Mining Industry on the Witwatersrand, 1902–1910', University of South Africa PhD Thesis, pp.144–149.

Harris, Karen (1998), 'A History of the Chinese in South Africa to 1912, D Litt et Phil', University of South Africa.

Jeeves, A. (1980), 'Het Volk and The Gold Mines: The Debate on Labour Policy, 1905–1910', University of the Witswatersrand, African studies seminar paper, 92.

McCallum, Simone Lisa (1994), 'Radical Racism in the Transvaal: F H P Creswell and the White Labour Movement, 1902–1912', Master Thesis, Queen's University, Kingston, Ontario, Canada

Reeves, J. A. (1954), 'Chinese Labour in South Africa, 1901–1910', MA thesis, University of the Witwatersrand.

Index

Abdurahman, Dr. Abdullah, 51, 191
Africans (native), 15, 18, 22–3, 25–9, 34, 36–7, 41, 43, 50–1, 53–5, 74, 76, 82, 92, 96, 98, 100–3, 114–19, 123–4, 126–7, 129, 131, 143, 149–51, 161, 170, 174–85, 177–9, 183
  see also Labour (African), Elections (1904 Cape)
Argus Group, the, *see* Newspapers
Afrikaners, 3–4, 7, 24, 27, 30, 34, 47–59, 103–8, 119, 130, 158, 165, 168–73, 178, 183, 195
  Cape Bond, 48–52, 55, 181
  Het Volk, 56, 165, 170–5, 177–8, 181–3, 191–2, 195
  Orangie Unie (ORC), 181
Anti-Asian associations, 44–6
  African Labour League, 46, 195, ch.3 ftnt.59
  National Democratic Federation, 46, 167, 195
  White League, 39, 46, 64, ch.3 ftnt.5 and ftnt.59
APO (African People's Organisation), *see* Abdurahman, Dr. Abdulla
Australia, 5, 9–10, 12, 14–16, 20–2, 25, 27, 30, 32, 35, 39–40, 44, 46, 59–63, 69, 103, 114, 144, 165, 173, 177, 186, 195, Introduction ftnt.24
  federation, 17–18, 41, 182
  NSW (New South Wales), 10, 12–13, 17–18, 20
  Northern Australia, 20
  Tasmania, 16
  Victoria, 13

Western Australia, 16, 20
  see also labour (white), Deakin, Alfred, Whiteside, Peter

Bailey, Abe, 31, 49, 174, 191, ch.2 ftnt.25, ch.3 ftnt.81 and 212, ch.7 ftnt.52
Blacks, *see* labour, African
Black Peril, 6–7, 96–9, 124, 179, 189
  see also Moral panics
Boers, *see* Afrikaners
Boer War, *see* South African War
Botha, Louis, 34, 55–7, 104–5, 170, 174, 179–80, 183, 191, ch.3 ftnt.129
  see also Het Volk, Afrikaners; Transvaal

Canada, 5, 14–15, 20–1, 40–1, 59–60, 62, 186
  British Columbia, 10, 12, 30, 60, 190
Cape Colony, 9, 23–4, 32, 41, 44–5, 47–55, 57, 59, 64, 66–7, 102, 130, 165, 176, 180, 182, 191, 193–5
  Afrikaner Bond, 48–52, 55, ch.3 ftnt.66; *see also* Afrikaners
  Cape Coloured, 51, 54–5, 124, 185
  Cape Town, 30, 48–9, 51, 181
  Progressive Party, 48–9, 51–3, 55, 57, 168, 181
  SAP (South African Party), 48, 51–2, 179, 181
  South African Liberal Association, 48, 165
  suffrage, 47, 51
  see also Election (1904 Cape)

Chamber of Mines, Transvaal, viii, 23, 27, 29–31, 34–6, 38, 61, 65, 67, 70, 82, 108, 117, 151, 156–7, 163, 178–9, 196
  Skinner Report, 35–6, 78, 114, 139
  CMLIA (Chamber of Mines Labour Importation Agency), 43, 56, 66–7, 79, 81, 84–5, 87–9, 91, 117, 123, 145, 147, 154–7, 190–1
Chamberlain, Joseph, 2, 19–21, 26, 33, 38, 41, 56, 65, 68–9, 83–4, 191
Chaplin, Francis 'Drummond' Percy, 43, 151, 153, 174, 191, 196
China
  1904 Labour Ordinance, 31, 36, 63, 67–70, 75–6, 78–9
  Boxer Rebellion (Boxers), 16, 18, 78, 80–1, 87, 95, 102, ch.4 ftnt.53, ch.5. ftnt.38
  Consul to Johannesburg, 81–2, 84, 110, 143, 148
  Consul to London, 80–2, 86
  languages, 7, 115–17
  migration, 3, 5, 8–21, 30, 32, 52, 59–60, 75, 86–8, 161–2, 182, 186–7, 189–90
  negotiations, over indenture, 79–84
  recruitment, 84–90
  see also Yellow Peril, Labour (Chinese)
Civilising mission, 7, 22, 52, 72
  missionaries, 29, ch.3 ftnt.37, ch.4 ftnt.126
  humanitarianism, 17, 71, 74, 187
Closer Union Society, 181
CMLIA (Chamber of Mines Labour Importation Agency), see Labour and Chamber of Mines
Colonial Conferences, 4, 16, 18, 20, 41, 64, 187
  colonial nationalism, 5, 39, 187, Introduction ftnt.20
  imperial conferences from 1907, 187–8
  see also Federation, Closer Union Society

Colonial Office, 11, 18–19, 24, 33, 37, 41, 47, 59–60, 65, 70, 77, 83, 94, 161, 185, 187–8
  see also Dominions Department
Colour bar, 76, 179
Corner House, 28–9, 31, 44, 58, 65, 77, 94, 152, 154–5, 169, 184, 178, 191–2, 194–5, ch.3. ftnt.211
Corporal punishment, see Violence
Corruption, accusations of, 64–7, 74, 113, 158, 172, 174, 190
  see also Economic imperialism, Newspapers
Churchill, Winston, 2, 191, ch.5 ftnt.54, ch.7 ftnts.48, 52 and 73
Creswell, Frederick, 28–9, 43, 74, 169–70, 174, 179, 192
crime, see Violence, Outrages

Deakin, Alfred, 59
  see also Australia
Dominions, 187–8

Eckstein & Co., see Corner House
Economic imperialism, 65, 71–2, 181, 189
  see also Corruption
Elections
  1904 Cape, 47–53, 55
  1906 British, 1, 68, 70, 72, 74, 146–7, 160–1, 165
  1907 Transvaal, 165, 167–74
  1910 South Africa, 171; see also Union, South African
ERPM (East Rand Proprietary Mines), 31, 85, 91, 93, 109, 152, 193
  see also Farrar, George
Evans, Samuel, 34, 169, 192–3
Evans, William (FLD Protector), 78–9, 83–4, 89, 120, 123, 144

Farrar, George, 31–6, 39, 43, 75, 84, 93, 109, 152–3, 164, 168–9, 174, 182, 193, 196, ch.2 ftnt.59
Federation (imperial and national), 5–6, 17, 38–9, 40–1, 46, 48, 59–60, 62, 69, 166–7, 170, 180–93, 187, 189, 192

Index

Fitzpatrick, James Percy, 26, 29–31, 168, 174, 191, 196
  see also Corner House
firearms, *see* Rearmament
FLD (Foreign Labour Department), 78, 83, 85, 87, 89–90, 96, 101, 108, 112–13, 116–17, 119, 123–4, 127–30, 143, 145–51, 154–67, 167, 177, 184, 194
  see also Evans, William, Jamieson, James
Fremantle, Henry Eardley Stephen, 48, 64, 193

Gandhi, Mohandas (Mahatma), 54, 193
  see also Indians

Homosexuality, *see* Labour (Chinese)
Hoover, Herbert, 88
Hull, Henry Charles, 174, 194

Imperial Conferences, *see* Colonial Conferences
Imperial Federation, *see* federation
Imperial South Africa Association (ISAA), 42, 66, 68
India Office, 19, 33, 71
Indians, 9–11, 15, 18–19, 21, 24, 33, 39, 41–2, 51, 54, 60, 71, 100–2, 103, 185–6, 188–9, 193
  Indian Opinion, 54, 102, 185, 193
Innes, James Rose, 120, 122, 130, 142, 194

Jabavu, John Tengo, 51, 194
Jameson, Leander Starr, 49–50, 53, 194
  Jameson Raid, 19, 31, 191, 193–5
Jamieson, James William (FLD Protector), 108, 120, 147, 150–1, 153, 155, 194
Japan, 6, 15, 18, 21, 44, 71, 80, 85, 123, 166
Jebb, Richard, 157
Jennings, Hennen, 30, 194

Labour Commission (1903), 32, 34–5, 196
Labour
  Chinese: crime, 100, 107, 127, 129–32, 147, 161, 169, 175, 186; disease, 15, 39, 50, 87, 148–9, 158, 161;gambling, 89, 95, 111, 113, 128, 130, 146–8, 154; police, 102, 104, 111–13, 115–17, 120–1, 124, 147–9, 153, 157–8; outrages, 96, 107–9, 120, 125, 129–30, 131, 142, 151, 156, 158, 162, 165, 171; secret societies, 145–6, 148; sex, 12, 15, 26, 34, 76, 96–9, 106, 113, 122–32, 153; uniforms, 102, ch.5 ftnt.35; *see also* China
  Indian, *see* Indians
  problem, the (or native question), 2, 8, 21–30, 35, 174–9
  white, 7, 12, 27–9, 34, 39, 45–7, 53, 61–3, 68, 71, 74–5, 99, 114, 167, 169–70, 174, 177–9, 183, 190
    see also Africans, Chamber of Mines, Creswell, Frederick, Legislation
Labour unions, 12–14, 16, 18, 27–9, 36, 43–7, 52, 66, 74, 99, 154, 161, 170, 173, 179–80, 189
Laurier, Wilfred, 20, 59, 187
Legislation, 13, 15–21, 31, 36, 39–41, 52–4, 64, 70, 75–6, 99, 148–9, 183, 185–6, 189
  see also Colour bar

Merriman, John Xavier, 48, 50–1, 69, 180, 182–3, 194
Milner, Alfred, 2, 23–6, 28–9, 31–3, 36, 38, 42, 47, 49, 55–62, 64–5, 67–9, 73, 79–80, 83–4, 142, 153, 168, 171, 177, 180–1, 184, 188, 192–6
Moral Panics, 6–7, 96–9, 103, 105, 107–8, 119, 122, 129–32, 143, 148, 155, 158, 165, 175, 179, 189, ch.5 ftnt.4
  see also Black Perils, Yellow Perils
Mozambique, 23, 26, 117, 123, 128, 175–6, 182

Index

Natal, 10, 15, 19, 24, 30, 33, 52, 60–1, 71, 96, 102, 180–2, 186
  Durban port, 60–1, 87, 110, 182
Natal Language Test, 20–1, 161
Newspapers, 11, 15, 17, 19, 21, 30, 36, 38, 42–4, 46, 50–1, 54, 60–1, 64–6, 69–70, 95–6, 98, 100, 102–3, 105, 108–10, 113–15, 119, 122–6, 129–30, 145, 154, 160, 165, 175, 186, 189
New Zealand, 16, 18, 20, 41, 46, 57, 63, 165–7, 186, 190

Opium, 10, 18, 40, 101, 113–14, 145–6, 148–9, 154
Orange River Colony, 32–3, 41, 58, 170, 180–2

Petitions, 21, 38, 43, 45, 47, 50–61, 63, 67, 84, 108, 113, 148, 160–1, 165–6, 172, 175, 182, 194–5
Philips, Lionel, 29, 31, 152–3, 174, 178, 195–6
Press, *see* newspapers

Quinn, J. W., 35, 195
  see also Labour Commission (1903)

Rearmament, 104–8, 171
Reconstruction, 6, 14, 24–5, 28
Responsible government, 19, 50, 61–2, 69, 162, 164, 166–8, 171, 174, 180
Robinson, Joseph Benjamin, 91, 94, 140, 187

Schumacher, Raymond, 195, ch.7 ftnt.58
Seddon, Richard, 20, 46, 59, 166, ch.7 ftnt.35
  see also New Zealand
Selborne, William Waldegrave Palmer, 2nd Earl of, 105, 107–8, 127–8, 142, 150–3, 156, 164, 180, 182, 194
Sex, *see* Labour (Chinese)

Sishuba, John Alf, 52, 195
Smuts, Jan C., 55–7, 104–5, 170, 173–4, 180, 182–7, 195, ch.3 ftnt.129
Solomon, Richard, 36, 79, 108, 127, 148, 151, 155, 157, 167, 172, 195
South African Union, *see* Union, South African
South African War, 1, 3–4, 7, 21–3, 26–7, 30–1, 41–2, 45, 48, 53, 56, 65–6, 68, 72, 74, 84, 160–1, 165, 190
  Vereeniging Treaty, 19, 33, 41, 51, 195
  War Contribution Loans, 26, 68, 99, ch.3 ftnt.202
Southern Rhodesia, 24, 32–3, 60, 96–7, 124, 140
Steyn, Marthinus T., 58, 105, 180–2
Sun Yat-Sen, 71, 81, 194

Tariff reform, 19, 41, 75, 191
  see also Chamberlain, Joseph
Transvaal
  Progressives, 168–9, 171–4, 181, 193
  Nationalists/Responsibles, 168, 172–4
  see also Afrikaners, Labour, Legislation, Responsible government, Elections, West Ridgway Committee

Union, South African, 180–3
United States, 5, 9, 11–15, 17, 19, 24, 30, 47, 62, 74, 81, 103, 121, 126, 163–4, 186–8, 195, ch.1 ftnt.39

Vereeniging Treaty, *see* South African War
Violence
  corporal punishment, 71, 82–4, 119–20, 144–5, 150–2, 194
  rape, 96, 98, 106, 124–5, 129
  symbolic violence, 98–9, 108, 115, 126, 140
  see also Labour (Chinese)

West Ridgeway Committee, 172–3
Wernher, Beit, & Co., 28, 30–1, 91, 93–4, 152, 174, 191–2, 195
  see also Corner House
White labourism, *see* Labour (white)
Whiteside, Peter, 35, 44, 195
  see also Australia

WNLA (Witwatersrand Native Labour Association), 23–4, 30, 35, 81–2, 84, 176–7, 182
  see also Labour
Wybergh, Wilfred John, 58, 169, 189, 196

Yellow Peril, 5, 7, 16, 18, 40–1, 47, 58, 73, 96–7, 98, 102–3, 124, 186, 189
  see also Moral Panics

Printed and bound by CPI Group (UK) Ltd, Croydon, CR0 4YY